Clara Pfäffle

Functional Imaging of Retinal Neurons

-

Phase-Sensitive Optical Coherence Tomography

Imprint of Infinite Science GmbH,
Technikzentrum | MFC 1
Maria-Goeppert-Straße 1
23562 Lübeck, Germany

Cover Design and Illustration: Infinite Science Publishing
Editorial and Copy Editing: Infinite Science Publishing

Publisher: Infinite Science GmbH, Lübeck, www.infinite-science.de
Printed in Norderstedt, Germany

ISBN: 978-3-945954-68-3

Bibliografische Information der Deutschen Nationalbibliothek:
Die Deutsche Nationalbibliothek verzeichnet diese Publikation in der Deutschen Nationalbibliografie; detaillierte bibliografische Daten sind im Internet über http://dnb.d-nb.de abrufbar.

Author

Dr. Clara Pfäffle
E-mail: cl.pfaeffle@uni-luebeck.de

Clara Pfäffle

Functional Imaging of Retinal Neurons

-

Phase-Sensitive Optical Coherence Tomography

Research Series of the
Institute of Biomedical Optics
University of Lübeck – Volume 3

Infinite Science
Publishing

Zusammenfassung

Die funktionelle Bildgebung von menschlichen Neuronen ist sowohl aus der Sicht der Forschung als auch der klinischen Diagnostik von großem Interesse. Dabei kann das Auge als optisches Fenster zum zentralen Nervensystem (ZNS) genutzt werden, da es anders als der Rest des ZNS nicht vom Schädelknochen abgeschirmt ist und somit für optische Bildgebungsmethoden zugänglich ist. Durch die Bildgebung mit sichtbarem oder nahinfrarotem Licht kann eine zelluläre Auflösung von wenigen Mikrometern in der Retina erreicht werden. Daher kann die Retina als Modellsystem genutzt werden, um in-vivo Untersuchungen des ZNS mit einer deutlich höheren Auflösung zu erzielen.

Die optische Kohärenztomographie (engl. optical coherence tomography, OCT) ist eine der Standardtechniken für die ophthalmologische Untersuchung. Sie ist in der Lage durch interferometrische Detektion des Tiefenprofils volumetrische Aufnahmen der Retina zu generieren. Dadurch ist die OCT prinzipiell in der Lage funktionelle Veränderungen in allen drei Raumrichtungen auf wenige Mikrometer genau ihrem Ursprungsort zuzuordnen.

Die interferometrische Messung ermöglicht es dabei, die Amplitude und die Phase des zurückgestreuten Wellenfeldes zu detektieren, wobei für die Darstellung der OCT Bilder üblicherweise nur die Amplitude verwendet wird, obwohl die Phase sehr viel sensitiver auf Veränderungen der Probe reagiert. Grund für diese Vernachlässigung der Phase ist, dass sie gerade auf Grund ihrer Sensitivität durch Bewegungen des Auges oder durch ein unpräzises Scannen stark von Artefakten überlagert ist, welche eine Interpretation des morphologischen Beitrags zur Phase in den meisten Fällen unmöglich machen.

In dieser Arbeit wird ein funktioneller Kontrast auf Grundlage der Phaseninformation des OCT Signals vorgestellt. Um Bewegungsartefakte zu verhindern, die während der Aufnahme eines Volumens auftreten, wurde sowohl die axiale als auch die laterale Datenerfassung parallelisiert. Dazu wurde ein Full-Field Swept-Source OCT (FF-SS-OCT) System verwendet. Durch die Parallelisierung in allen drei Raumrichtungen sind die Phasen der unterschiedlichen Datenpunkte innerhalb eines Volumens nicht mit Bewegungsartefakten überlagert. Die Volumina sind sozusagen phasenstarr. Um diese Phaseninformationen jedoch für eine funktionelle Auswertung nutzen zu können, ist ein weitere Nachbearbeitung der Daten notwendig. Um die zeitlichen Veränderungen der Phase zu messen, müssen die Bewegung zwischen aufeinanderfolgenden Volumen entfernt werden. Dies wird durch eine Koregistrierung der Volumendaten sichergestellt, welche sowohl Translation, Rotation als auch Deformation in allen drei Raumrichtungen berücksichtigt. Weiterhin müssen die zu untersuchenden Strukturen in der Retina, reproduzierbar segmentiert werden. Um den Anforderungen eines hohem Rauschpegels, sowie den großen dreidimensionalen Datensätzen gerecht zu werden, wurde an dieser Stelle ein Segmentierungsalgorithmus verwendet, welcher auf einem 3D-graphenbasierten Ansatz beruht und mit einem Clusterverfahren kombiniert wurde. Zum Schluss wurde eine spezielle Phasenauswertung verwendet, welche die statistischen Eigenschaften von zufälligen und systematischen Phasenänderungen nutzt, um die Phasenveränderung über längere Zeiten hinweg sinnvoll auszuwerten.

Der phasenbasierte funktionelle Kontrast wurde an der Retina eines freiwilligen Probanden evaluiert. Dabei konnte funktionelle Veränderung in der Phase des in den Photorezeptoraußensegmenten (AS) und der inneren Plexiformschicht (engl. inner plexiform layer, IPL) gestreuten Lichts nach optischer Stimulation gemessen werden. Die hohe Sensitivität der Phase ermöglicht dabei, morphologische Veränderungen mit einer Genauigkeit von wenigen Nanometern zu detektieren. Die gemessenen Zeitverläufe der funktionellen Änderungen sind dabei höchst reproduzierbar und zeigen eine Verlängerung der optischen Weglänge von mehreren hundert Nanometern in den Photorezeptoren. Die funktionellen Veränderungen wurden in Abhängigkeit von verschiedenen Stimulierungsparametern untersucht. Weiter konnte die funktionelle Veränderung einzelnen Photorezeptoren zugeordnet werden. Die dreidimensionale Bildgebung ermöglichte die unterschiedlichen Tiefenebenen der Retina in einer einzigen Messung zu untersuchen und funktionelle Signale in verschiedenen Zellebenen gleichzeitig zu detektieren. Durch die simultane Messung der funktionellen Veränderungen in den Photorezeptoren und der IPL, konnte die Verschaltungsarchitektur der Photorezeptoren mit den nachgeschalteten Bipolar- und Ganglienzellen untersucht werden.

Durch eine Tiefendiskriminierung des Signals innerhalb der IPL konnten zudem Beiträge von unterschiedlichen Zelltypen differenziert werden, welche verschiedene Aspekte des visuellen Signals verarbeiten. Dadurch konnte gezeigt werden, dass Zelltypen die zum Signal in der Mitte der IPL beitragen, sensitiv auf zeitliche Veränderung im visuellen Signal reagieren.

Die in dieser Arbeit gezeigten Ergebnisse, demonstrieren somit das Potential der phasensensitiven OCT für die funktionelle Bildgebung von retinalen Neuronen. Dabei ermöglicht vor allem die Differenzierung der funktionellen Veränderung in allen drei räumlichen Dimensionen eine Diskriminierung der Beiträge von unterschiedlichen Zellklassen, wodurch sich neue Möglichkeiten für die funktionelle Untersuchung der Retina in-vivo eröffnen.

Abstract

Functional imaging of human neurons is of great interest from the perspective of both, research and clinical diagnostics. For imaging, the eye can be used as an optical window to the central nervous system (CNS), since, unlike the rest of the CNS, it is not shielded by the skull and is therefore accessible by optical imaging methods. By imaging with visible or near-infrared light, resolution of a few micrometers in the retina can be achieved. Therefore, the retina can be used as a model system to achieve in-vivo examinations of the CNS with cellular resolution.

Optical coherence tomography (OCT) is one of the standard techniques for ophthalmological examinations. It can generate volumetric images of the retina by interferometric detection of the depth profile. Thus, OCT is in principle able to identify functional changes in all three spatial directions with an accuracy of a few micrometers. The interferometric measurement allows the detection of the amplitude and phase of the backscattered radiation. Usually, only the amplitude is used for the generation of OCT images, although the phase provides very sensitive information on changes in the sample. The reason for this neglect of the phase is that it is overlaid with huge artifacts due to eye movement or imprecise scanning, which makes it virtually impossible to interpret the phase, in most cases.

In this thesis, a functional contrast based on the phase information of the OCT signal is presented. To avoid motion artifacts that occur during volumetric imaging, both axial and lateral data acquisition was parallelized. To do this, a full-field swept-source OCT (FF-SS-OCT) system was used. Such parallelization in all three spatial directions ensures that the phases of the different data points within a volume are not overlaid with random motion artifacts. Such volumes are called phase-locked. However, in order to use this phase information for a functional evaluation, further, post-processing is necessary. To measure the temporal changes of the phase, the motion between successive volumes must be removed. This is ensured by a co-registration of the volume data, which considers translations, rotations, and deformations in all three spatial directions. Furthermore, the retinal structures to be examined for functional changes must be reproducibly segmented. To segment despite high-noise and three-dimensional data sets, a segmentation algorithm based on a 3D graph-based approach was used and combined with a clustering method. Finally, a dedicated phase evaluation was used, which uses the statistical properties of random and systematic phase changes to evaluate the phase change over longer periods in a meaningful way. In order to evaluate the physiological significance of the observed changes in phase, the retina of a volunteer was examined. Changes in the phase of the radiation scattered by the photoreceptor outer segments (OS) and the inner plexiform layer (IPL) were measured before, during, and after optical stimulation. The high sensitivity of the phase allowed the detection of morphological changes with an accuracy of a few nanometers. The measured time courses of the functional changes were highly reproducible and showed an increase in the optical path length of several hundred nanometers in the photoreceptors. These functional changes were investigated for different stimulation parameters. Furthermore, functional changes could be assigned to individual photoreceptors. Due to the three-dimensional spatial resolution of the imaging, it was possible to investigate different depth levels of the retina simultaneously, which allows simultaneous detection of functional signals in different cell levels. By recording the functional changes in photoreceptors and IPL at the same time, the interconnection architecture of photoreceptors with the downstream bipolar and ganglion cells could be investigated

By depth discrimination of the signal in the IPL, contributions of different cell types sensitive to different aspects of the visual stimulus could be differentiated. Thus, it could be shown that cell types that contribute to the signal in the middle of the IPL are especially sensitive to temporal changes in the visual signal.

The results shown in this thesis demonstrate the potential of phase-sensitive OCT for functional imaging in the neuronal retina. In particular, the 3D-spatially resolved differentiation of the functional alteration of cells allows the discrimination of the contributions of different cell classes, thus opening up new possibilities for functional imaging of the retina in vivo.

Contents

List of Figures

List of Tables

Abbreviations

Abbreviation	Meaning
AO	adaptive optics
AOSLO	adaptive optics scanning laser ophthalmoscopy
AMD	age-related macular degeneration
ATP	adenosine triphosphate
BBB	blood-brain barrier
cGMP	cyclic guanosine monophosphate
CNG	cyclic nucleotide-gated
CNS	central nervous system
CT	computer tomography
DSGC	directionally selective ganglion cells
FAD	flavin adenine dinucleotide
FD-OCT	Fourier-domain optical coherence tomography
FF-OCT	full-field optical coherence tomography
FF-SS-OCT	full-field swept-source optical coherence tomography
FF-TD-OCT	full-field time-domain optical coherence tomography
FWHM	full width at half maximum
fMRI	functional magnetic resonance imaging
GCL	ganglion cell layer
GDP	guanosine diphosphate
GTP	guanosine triphosphate
INL	inner nuclear layer
IPL	inner plexiform layer
IS/OS	inner segement/outer segment junction
LED-GC	local edge detector ganglion cell
MRI	magnetic resonance imaging
MS	multiple sclerosis
NADH	nicotinamide adienine dinucleotide
NFL	nerve fiber layer
NIR	near infraread
OCT	optical coherence tomography

OPL	outer plexiform layer
OS	outer segment
PD	Parkinson's disease
PDE	phosphodiesterase
PNS	peripheral nervous system
PSF	point-spread function
R	rhodopsin
RPE	retinal pigment epithelium
SD-OCT	spectral-domain optical coherence tomography
SLO	scanning laser ophthalmoscopy
SNR	signal-to-noise ratio
SS-OCT	swept-source optical coherence tomography
SVM	support vector machine
T	transducin
TD-OCT	time-domain optical coherence tomography

1 Introduction

The wiring and communication of single neurons in the central nervous system (CNS) is a research area of tremendous interest. By observing the neuronal behavior one can not only learn about the general working principle of the brain but it can also be of advantage for early diagnostic and treatment monitoring of neuronal degenerative diseases. For this purpose, an imaging modality is needed that is able to image morphology and function of in-vivo tissue of the CNS with a cellular resolution. Standard imaging modalities for the CNS, like computer tomography (CT) and magnetic resonance imaging (MRI), enable non-invasive imaging of the CNS in humans but they lack a functional contrast. Functional MRI (fMRI) offers functional contrast by measuring changes in blood oxygen level in activated brain areas. However, the centimeter resolution of such images are far away from resolving single neurons.

Optical imaging with wavelengths between 400 nm and 1 µm enables a resolution on the cellular level. However, light in the visible and near-infrared (NIR) does not readily penetrate the skull undisturbed for imaging brain tissue.

1.1 The Retina as a Model System

The eye is the only site, where the CNS is directly accessible for optical imaging. Due to its design, the retina can be imaged with visible and NIR light. Additionally, the retina is the only sensory system in the human body, that evolves from the diencephalon during embryonic development and can therefore be considered as a part of the CNS [1]. In fact, the retina shares many similarities with the brain. It exhibits a specialized immune response and a blood-retina barrier, which can be considered as a counterpart of the blood-brain-barrier [2, 3]. Given these similarities of retina and brain, one can hope to learn about the function and pathology of the brain from the retina. In fact, many CNS disorders correlate with functional impairment and morphological changes in the eye.

For example, in most patients who suffer from multiple sclerosis (MS), visual impairments develop during the disease and in 50% of the patients, visual loss is a present symptom [4–6]. In 75% of MS patients, optic neuritis is diagnosed, which goes along with degeneration and in some cases atrophy of retinal ganglion cells and axons [7], as well as a thinning of the retinal nerve fiber layer [8, 9]. A thinning of the retinal nerve fiber layer is also observed in Parkinson's disease (PD), where it coincides with a swelling of the retinal ganglion cells and photoreceptor cells [10–12]. In Alzheimer's disease retinal ganglion cell atrophy and reduced blood flow have been observed [13–16]. These morphological changes go along with functional impairment in vision, like reduced visual acuity and contrast sensitivity or disturbed color vision [8, 12, 14–16]. In some cases, those functional deficits are present without any detectable morphological changes.

Therefore, it is reasonable to assume that functional changes of the different retinal neurons would precede morphological changes and that we would reach a deeper understanding of the beginning of these diseases and gain the opportunity of an early diagnosis with functional imaging of the retina [1].

1.2 Functional Imaging of the Retina

Imaging the Blood Flow To image functional changes in the retina, a contrast parameter is needed that indicates the transition to an activated state of the different cells in the neuronal tissue. The high metabolic demands, which are connected with neuronal activity, makes the blood flow a good parameter for this task. Techniques like adaptive optics scanning laser ophthalmoscopy (AOSLO) can measure the velocity of single blood cells [17–19] and were able to show that the velocity of those blood cells increases in larger arterioles when stimulated with light [20]. Besides AOSLO, the Doppler shift that is caused by the blood flow in interferometric techniques such as laser Doppler, laser Doppler holography, or optical coherence tomography can be used to determine the velocity of blood cells in larger vessels of the retina [21–27]. Using these techniques an increase in blood flow was also shown, when stimulating the retina with flickering light [28, 29].

Additionally, absorption changes in the visible and NIR range can provide metabolic contrast. For example, the absorption spectrum of hemoglobin changes when binding oxygen [30]. In oximetry, this is used to measure the blood-oxygen-level, which has a saturation of 35-43% in retinal circulation [31, 32]. Thus, optical oximetry yields similar information about the neurovascular coupling as fMRI, i.e., it shows an increase in blood flow that takes place to cover the increased demands for oxygen and glucose in neurally active tissue, however, with a higher temporal and spatial resolution.

Imaging Spectral Changes Although changes in the blood flow velocity and oxygenation level provide functional contrast, they do not provide cellular resolution. Besides hemoglobin, other molecules that participate in cellular metabolism or signal transduction change their optical properties as well when they are activated. One of them is the retinal photopigment rhodopsin, situated in the photoreceptor outer segments. It absorbs photons to initiate the visual signal transduction. Once a molecule has absorbed a photon, it changes its conformation and can therefore not be activated by further photons until it is recovered. Such photopigments are bleached, thus causing changes in the absorption spectra of visible light as well. This effect is used in rhodopsin densitometry [33]. It can be used to measure the photosensitivity [34, 35] or the rate of rhodopsin regeneration [36]. Combining the densitometry with adaptive optical (AO) imaging, single photoreceptor cells can be resolved and analyzed separately, which enables cone classification [37, 38]. Although those opsin densitometry measurements have given a massive insight into the physiology of the photopigments and are able to image functional changes in single cells, the measurements are time-consuming and they are limited to functional imaging of a single molecular pathway in the photoreceptor cells.

Fluorescence Imaging Besides functional changes in the absorption spectra, functional imaging can also be done via fluorescent molecules. Such molecules usually contain double bonds. Through light absorption, the π-electrons of those double bonds can be brought into an excited singlet state, emitting light at a longer wavelength when they return to their ground state. These fluorescence markers can be used to label and track molecular processes and structures. For example, neuronal activation can be visualized by calcium indicators, voltage sensors, and glutamate markers [39]. Since extrinsic fluorescence biomarkers have to be brought into the organism, those images can so far only be done ex-vivo or in animal studies. However, some fluorescent molecules are naturally present in the cells such as nicotinamide adenine dinucleotide (NADH) and flavin adenine dinucleotide (FAD). These molecules are involved in the glycolysis, citric acid cycle, and oxidative phosphorylation. Those pathways are responsible for adenosine triphosphate (ATP) production and should therefore increase when there is a higher metabolism in the cells. Flavoprotein and NADH autofluorescence imaging has been reported for in vivo retina [40,41] and varying fluorescence was attributed to disease progression [42] but so far it has not been studied in response to visual stimulation. Apart from these fluorescence molecules, which are common in all cells, more specialized pathways, like the visual cycle, can also be imaged by fluorescence imaging using retinoid as a target molecule. Thereby the retinoid production, clearance, and the establishment of an equilibrium in response to light adaption could be shown in primates [43].

Imaging Changes in the Scattering Behavior A completely different approach is to measure changes in tissue scattering. Unlike the above mentioned methods it is not possible to assign such changes to a specific molecular process, since they can be caused by very different processes. Nevertheless, this method also has some advantages. For example, it can be used with light in the NIR-range, which, firstly, poses no photochemical hazard for the retina, secondly, no significant stimulation takes place through the measurement itself, and, thirdly, such measurements can easily be performed with imaging modalities that have already found their way into ophthalmological diagnostics, such as scanning laser ophthalmoscopy (SLO) and OCT.

Using SLO with 900 nm wavelength, a decrease of backscattered light could be detected after visual stimulation in primates [44] and humans [45]. Those changes in backscattered light arose only in the stimulated area [46]. However, studies in different animals show very different responses to visual stimulation. While imaging of macaque, monkey, and human show primary a decrease in backscattered light [46–48], in other animals like cats an increase as well as a decrease of backscattered light in response to light were observed for different locations [49], and measurements in rabbits show a biphasic time course of the back scattered light [50].

Combining SLO with AO, such intensity changes could be correlated with the different neuronal cells in the human retina. The main contribution thereby came from the photoreceptor outer segments [51]. Further research using AOSLO was able to obtain the backscattered light changes for single cone outer segments exhibiting both light decrease and increase in several degrees for human photoreceptors [52], suggesting that the previously measured changes using SLO were some kind of ensemble average. It is believed that these changes are caused by interference of the two ends of the photoreceptor outer segments [53]. The changes in intensity would therefore be correlated to a length change of the outer segments. Whether an increase or a decrease of the backscattered light is observed, therefore very sensitively depends on the average length of the outer segments and the used wavelength.

Signal fluctuations upon optical stimulation were also observed with OCT. First, those changes were shown in ex vivo experiments of rabbit retina [54] and shortly after that in humans [55]. Due to the tomographic imaging of OCT, changes in backscattered light can not only be observed in a single neuronal layer, but it is possible to observe functional changes of different neuronal layers simultaneously. This could not only enable the monitoring of different neurons with a single imaging modality but could also allow us to investigate the wiring between different neuronal layers. In fact, functional imaging based on intensity fluctuation of light reflected in the inner retinal layers like the inner and outer plexiform layers (IPL and OPL) have already been shown in anesthetized animals [56, 57].

However, using OCT the observed intensity changes are not caused by direct interference of the front and back of the photoreceptor, rather it is believed that in this case they are caused by a change of speckle intensity, which in turn are caused by morphological changes below the resolution of the system and therefore could still have the same origin as the intensity fluctuation observed with SLO. However, those fluctuations in intensity observed with OCT were, similar to the ones measured with SLO, very noisy and difficult to reproduce. In order to use such measurements for diagnostic or medical trials, the changes have to be measured much more sensitively.

1.3 Phase Sensitive Imaging

An optical parameter that has not been mentioned so far is the phase of the detected light. Detecting the phase of a wavefront is a very powerful tool. It carries information about the sample structure on a sub-wavelength scale, which enables detecting changes in a single-digit nanometer range. The phase should therefore be much more sensitive to changes in the depth profile than the amplitude of the backscattered light.

There have already been in vitro studies showing that small morphological changes, which occur in the plasma membrane of a neuron during its excitation can be measured with phase-sensitive methods [58–61]. Unfortunately, the optical phase is in most imaging modalities not known, since in general only the intensity of a wave field can be directly measured. In coherent, interferometric imaging modalities such as OCT, the phase, however, can be extracted from the intensity data. Thus, using OCT, one has in general access to the phase of the backscattered wave. However, the formation of speckles in coherent imaging complicates the interpretation and evaluation of the phases. Additionally, since the phase is very sensitive to small changes, retinal movement and measuring artifacts render it in most cases completely useless for further interpretation.

To avoid those disturbing influences, retinal volumes have to be imaged with very high velocities to create phase-locked volumes. Those phase-locked volumes can be acquired with a very fast scanning OCT system [62–65]. To realize such fast scanning there are high technical requirements on the used hardware components and their adjustment.

Another method to generate phase-locked OCT volumes is by using full-field swept-source OCT (FF-SS-OCT). This technique circumvents the lateral scanning of the sample by illuminating the field-of-view with a parallel beam and thereby parallelize the data acquisition. Compared to phase-locked, scanning OCT systems, FF-SS-OCT has little requirements on the hardware adjustment of the setup but needs extensive post-processing to create competitive images.

However, FF-SS-OCT has already shown how powerful the detection of the phase of the backscattered wavefield could be. Besides numerical improvements of image quality by computational correction of dispersion artifacts [66, 67] and aberrations [68, 69], functional applications such as measuring the blood flow dynamics in the retina [70] or detecting three-dimensional micromotion were shown [70].

1.4 Structure of this Thesis

In this thesis, functional imaging of the human retina is realized with phase-locked FF-SS-OCT.

In the following chapter, the physical principles for phase-sensitive imaging are explained. In particular, the technique of OCT is discussed. In addition, this chapter provides the biological basics of the retina and its neurons.

In Chapter 3 a detailed description of the utilized FF-SS-OCT setup is given. Furthermore, the post-processing for the evaluation of the phase is explained. To ensure a phase-sensitive, functional analysis of the retina, different steps have to be done in the post-processing, including the correction of present phase errors, a co-registration of all volumes in one measurement to remove the global motion in between adjacent volumes, a segmentation of the structure of interest, and a specialized phase evaluation, that considers the speckle formation.This specialized FF-SS-OCT setup combined with sophisticated post-processing enables the detection of phase-locked measurements over up to 8 seconds, which can be used for functional imaging in the retina.

In Chapter 4 the results of functional imaging in the living human are presented. By using phase-sensitive evaluation, the elongation of the photoreceptor outer segments could be measured with a nm accuracy. Additionally, functional changes could also be measured in the inner plexiform layer. Both functional signals exhibited a highly reproducible time-course and it was possible to detect small changes down to a single-digit nanometer range. The functional signals were analyzed regarding its spatial appearance, as well as its dependency regarding several stimulus parameters.

In Chapter 5 the results are discussed, first, by comparing them to other phase-sensitive measurements and, in a second part, based on the results shown in Chapter 4, Different explanations regarding the physiological origin are discussed.

This chapter is followed by a conclusion and an outlook for possible future work, regarding an improvement of the FF-SS-OCT setup, an exploration of alternative imaging methods for phase-sensitive imaging in the retina, further experiments for determination of the molecular origin of this signal, and the evaluation of the clinical relevance of the shown functional signals.

2 Theory

In the following chapter, the mathematical, physical, and biological background necessary to follow this work will be explained. For the numerical reconstruction of the image information, the Fourier transform is of central importance. Therefore, first, its mathematics and relevant properties are explained, as well as its significance for optics. Additionally, analytic signals and their significance for phase-sensitive imaging are presented. This is followed by the basics of optical coherence tomography (OCT), its signal generation, and the formation of speckles. Especially, full-field swept-source OCT (FF-SS-OCT), which is used in this thesis, is discussed. Additionally, the origin and influences of phase errors still present in phase-locked imaging are reviewed.

The chapter is concluded by an introduction to the biology of the nervous system and the retina, starting with a rough overview, followed by a more detailed insight into the different retinal cell types, as well as their structure and function.

2.1 Fourier-Transformation in Signal Processing

The Fourier-transform is a mathematical operation that is used in a wide range of scientific and technical applications. Also, in the context of this work, it is used both in the reconstruction of Fourier-domain OCT, as well as in its post-processing.

The Fourier-transform is an unambiguous integral transform. Such operations transfer integrable functions, which map a function from a variable space $x \in X$ onto a new variable space $k \in K$. They are completely lossless and in the case of the Fourier transform reversible. That means, that the whole information of the original function is preserved, but is represented in a new basis. The transition of a function to a new variable space can be advantageous, since certain mathematical operation may be computationally less complex.

The Fourier-transform in particular deconstructs the original function $f(x, y)$ in periodic signals with different amplitudes, frequencies, and phases. Thereby the function is transferred from a spatial domain (x, y) into a frequency (k_x, k_y) domain. The Fourier-transform of a function $f(x, y)$ into a function $\tilde{f}(k_x, k_y)$ is defined by:

$$\mathcal{F}\{f(x,y)\} = \tilde{f}(k_x,k_y) = \iint\limits_{-\infty}^{+\infty} \mathrm{d}x\,\mathrm{d}y\; f(x,y)(\cos(k_x x + k_y y) - \mathrm{i}\sin(k_x x + k_y y)) \quad (2.1)$$

$$= \iint\limits_{-\infty}^{+\infty} \mathrm{d}x\,\mathrm{d}y\; f(x,y)e^{-\mathrm{i}(k_x x + k_y y)} \quad (2.2)$$

After a Fourier-transform not only certain mathematical operations are computationally less complex but also analysis and manipulation of specific parameters, which are not accessible in the spatial domain, become possible. For full use of the Fourier-transform, it is important to know its properties, which are listed below, following the presentation in 'Fourier Optics' by Joseph W. Goodman [71].

- **Linearity**

$$\mathcal{F}\{\alpha g + \beta h\} = \alpha\mathcal{F}\{g\} + \beta\mathcal{F}\{h\}$$

The linearity signifies that the Fourier-transform of a weighted sum of two (or more) functions is the identically weighted sum of their individual transforms.

- **Fourier-Shift Theorem** If $\mathcal{F}\{g(x,y)\} = \tilde{g}(k_x,k_y)$ then

$$\mathcal{F}\{g(x-a,y-b)\} = \tilde{g}(k_x,k_y)e^{-\mathrm{i}(k_x a + k_y b)}.$$

Since the image information is not changed by a shift in the spatial domain the spectral decomposition does not change as well. A translation in the space domain introduces only a linear phase shift in the frequency domain.

- **Parseval's Theorem** If $\mathcal{F}\{g(x,y)\} = \tilde{g}(k_x,k_y)$ then

$$\iint\limits_{-\infty}^{\infty} \mathrm{d}x\,\mathrm{d}y |g(x,y)|^2 = \frac{1}{(2\pi)^2} \iint\limits_{-\infty}^{\infty} \mathrm{d}k_x\,\mathrm{d}k_y\, |\tilde{g}(k_x,k_y)|^2.$$

The integral of the left-hand side can be interpreted as the energy of the waveform $g(x,y)$. Due to the energy conversation this leads to the energy density in the frequency domain $|\tilde{g}(k_x,k_y)|^2$.

- **Convolution Theorem** If $\mathcal{F}\{g(x,y)\} = \tilde{g}(k_x,k_y)$ and $\mathcal{F}\{h(x,y)\} = \tilde{h}(k_x,k_y)$ then

$$\mathcal{F}\left\{\iint\limits_{-\infty}^{\infty} \mathrm{d}\xi\,\mathrm{d}\eta\, g(\xi,\eta)h(x-\xi,y-\eta)\right\} = \tilde{g}(k_x,k_y)\tilde{h}(k_x,k_y).$$

The convolution of two functions in the spatial domain is entirely equivalent to multiplying their individual transforms and inverse transforming. This is especially important when huge data sets should be convoluted in order to filter for example edges, high or low frequencies or to create a soft-focus effect by a convolution with a Gaussian kernel.

- **Complex Conjugation** If $\mathcal{F}\{g(x,y)\} = \tilde{g}(k_x,k_y)$ then

$$\mathcal{F}\{g^*(x,y)\} = \tilde{g}^*(-k_x,-k_y).$$

The Fourier-transformation of a complex conjugated function corresponds to the mirrored and complex conjugated Fourier-transformation of the original function.

- **Auto-correlation** If $\mathcal{F}\{g(x,y)\} = \tilde{g}(k_x,k_y)$ then

$$\mathcal{F}\left\{\iint\limits_{-\infty}^{+\infty} \mathrm{d}\xi\,\mathrm{d}\eta\, g(\xi,\eta)g^*(x-\xi,y-\eta)\right\} = |\tilde{g}(k_x,k_y)|^2.$$

Considering the previous two theorems it becomes obvious that an auto-correlation of a function can be described by the squared magnitude of its transformation.

- **Similarity** If $\mathcal{F}\{g(x,y)\} = \tilde{g}(k_x,k_y)$ then

$$\mathcal{F}\{g(ax,by)\} = \frac{1}{|ab|}\tilde{g}\left(\frac{k_x}{a},\frac{k_y}{b}\right).$$

The similarity of the Fourier-transformation states that a compression of our function in the spatial domain corresponds with a dilation in frequency domain.

2.2 Fourier-Transformation in Optics

Besides the many benefits the Fourier-transform provides for signal processing, it is also of high significance in other areas. One of them is optics. This becomes obvious when the mathematical representation of electro-magnetic waves is considered. In general lightwaves can be described by a solution of the wave equation:

$$\nabla^2 U(\boldsymbol{x},t) - \frac{n^2}{c_0^2}\frac{\partial^2}{\partial t^2}U(\boldsymbol{x},t) = 0$$

Here $U(x,t)$ denotes the space and time-dependent scalar wavefield, c_0 is the vacuum light velocity, and ∇^2 denotes the Laplace operator. For the simplified assumption of monochromatic light,

$$U(\boldsymbol{x},t) = A(\boldsymbol{x})\sin(\boldsymbol{k}\cdot\boldsymbol{x} - \omega t)$$

represents a harmonic solution of the wave equation [72]. Here $A(x)$ describes the spatial dependent amplitude, ω the angular frequency, and $\boldsymbol{k} = n\frac{\omega}{c_0}$ the wavenumber of the wavefield. Because of its linearity, the linear combination of all kinds of solution are again solutions of the wave equation:

$$U(\boldsymbol{x},t) = aU_1(\boldsymbol{x},t) + bU_2(\boldsymbol{x},t),$$

with a and b being arbitrary constants. With this equation, it becomes obvious that a wavefield can be completely described by decomposition into periodic signals. Just like a Fourier-transformation $\mathcal{F}\{U(\boldsymbol{x},t)\} = \tilde{U}(\boldsymbol{k},t)$ would do. And indeed, an electro-magnetic wave field can be observed in optical phenomena represented onto a basis of its spatial variables $U(\boldsymbol{x},t)$ as well as onto a basis of its frequencies variables $\tilde{U}(\boldsymbol{k},t)$, like the following examples should illustrate.

One frequently used and simple example of Fourier optics is a converging lens. Such a lens multiplies the wavefield of the incident light with a spatially dependent phase factor, in general, used to focus the light scattered by an object onto an image plane to recreate the image of that object. However, for the wave field which is in the positive focus position of the lens, this phase factor will create a Fourier-transform of the wavefield in the negative focus position of the lens. Thus, such lenses can be used as well to manipulate a wave field in the frequency domain, similar to the mathematical operation in signal processing. This is, for example, used in darkfield microscopy and some techniques of phase-contrast imaging [73].

Another example of the Fourier-transformation in optics is the diffraction of electromagnetic waves. In the farfield (Fraunhofer region), the wave field of a diffraction pattern generated by a slit can be described by a sinc-function, which is the exact Fourier-transform of a rectangular function, which the slit resembles. According to the similarity property of the Fourier-transform, the sinc-function will extend, when the slit becomes smaller, thereby the secondary maxima move away from the center. Thus conclusions about the size of the slit can be drawn, even if we are not able to measure it directly. This relation is for instance used in X-ray diffraction analysis to determine the atomic structure of crystalline solids using its diffraction pattern [74].

These two examples illustrate that the propagation of a wavefield is deterministic, and therefore can be completely described by mathematical operations. This motivates the research area of computational optics. If the complete mathematical description of a wavefield at a certain time is given with both amplitude and phase, then there is no need for optical instruments anymore. The propagation of that wavefield can be done numerically, as well as adding artificial phase factors in order to refocus an image or to remove distortions of the wave field. Thereby, complex adjustments and expensive components can be avoided.

The general problem for computational optics is that the wavefield of visible or near-infrared light cannot be detected directly. Instead only the intensity $I = \langle U^*(\boldsymbol{x},t)U(\boldsymbol{x},t)\rangle_t = \langle |U(\boldsymbol{x},t)|^2\rangle_t$ of a wavefield can be detected. Here, $\langle\rangle_t$ denotes the temporal average of the wave field. During the detection of the intensity, information gets lost. In incoherent imaging a complete mathematical description of the wavefield is not possible, and therefore the numerical propagation and manipulation of the wavefield neither. However, in coherent imaging modalities, it is possible to retrieve the whole information from the detected intensity, if the detected signal corresponds to a so-called analytic signal. An analytic signal is a complex-valued function whose spectrum contains no negative frequencies.

2.3 Phase Detection, the Analytic Signal and the Hilbert-Transform

In the following section, it will be demonstrated which information of the complex wavefield is lost by the detection of the intensity and how it can be recovered if the detected signal is analytic.

For this purpose, it is assumed that one is interested in detecting the full information of the superposition of two monochromatic wave fields $U_1(z)$ and $U_2(z)$ of the same light source. The intensity I_{sup} of this superposition is then given by,

$$I_{\text{sup}} = \left\langle |U_1 + U_2|^2 \right\rangle_t \tag{2.3}$$

$$= \langle U_1^* U_1\rangle_t + \langle U_2^* U_2\rangle_t + \langle U_1^* U_2\rangle_t + \langle U_1 U_2^*\rangle_t. \tag{2.4}$$

Here, the first two terms correspond to the intensity of the wavefield U_1 and U_2, respectively, while the last two terms describe the interference term of these two wavefields including their phase according to their optical path length difference. In the following one is only interested in this interference term of the two waves. Therefore, $S = U_1^* U_2$ is introduced. Thus, the information detected from this interference can be described by

$$2\mathfrak{Re}\{S\} = S^* + S = \langle U_1^* U_2 \rangle_t + \langle U_1 U_2^* \rangle_t. \tag{2.5}$$

Therefore, by detecting the intensity only the real part of the complex signal S can be detected, while the imaginary part and therefore the phase information is initially lost. Such an incomplete description of a wavefield can no longer be used for computational optics since essential information of the interference are missing.

For example, the Fourier-transform of such an incomplete description of a wavefield, does no longer yield the correct frequency spectrum of the complex Signal S. If only the real part of a complex function is Fourier-transformed, the complex spectrum $\tilde{S}(k)$ of the function is superimposed with its complex conjugated counterpart:

$$\mathcal{F}\{\mathfrak{Re}(S(z))\} = \frac{1}{2}[\mathcal{F}\{S^*(z)\} + \mathcal{F}\{S(z)\}] \tag{2.6}$$

$$= \frac{1}{2}(\tilde{S}^*(-k) + \tilde{S}(k)) \tag{2.7}$$

If $\tilde{S}(k)$ contains both negative and positive frequencies, then those two terms cannot be distinguished from one another. If, however, it is known that an analytic signal is observed, meaning that the interference signal does not contain any negative frequencies, i.e., $\tilde{S}(k) = 0$ for $k < 0$, then the complex conjugated counterpart only contains negative frequencies. Therefore the signal in the frequency domain can be separated from its complex conjugated counterpart, by setting all negative frequencies to zero and multiplying the positive frequencies with a factor of two to account for energy conservation. This is done by using the sign function, which is -1 for all $k < 0$ and 1 for all $k > 0$.

$$\tilde{S}(k) = (1 + \text{sign}(k))\mathcal{F}\{\mathfrak{Re}(S(z)\} \tag{2.8}$$

$$= \mathcal{F}\{\mathfrak{Re}(S(z))\} + \mathcal{F}\Big\{\mathcal{F}^{-1}\{\text{sign}(k)\} * \mathfrak{Re}(S(z))\Big\} \tag{2.9}$$

By separating those two terms the complete complex description of an analytic signal can be retrieved. With the Fourier-transform of the sign-function $\mathcal{F}\{\text{sign}(k)\} = {}^{\mathrm{i}}/_{\pi z}$, the complex signal in the spatial domain can then be described by

$$S(z) = \mathfrak{Re}(S(z)) + \mathrm{i}\frac{1}{\pi}\Big(\frac{1}{z} * \mathfrak{Re}(S(z))\Big) \tag{2.10}$$

$$= \mathfrak{Re}(S(z)) + \mathrm{i}\frac{1}{\pi}\int_{-\infty}^{+\infty} \mathrm{d}z' \, \frac{1}{z - z'}\mathfrak{Re}(S(z')), \tag{2.11}$$

where $*$ denotes a convolution. The second part of the equation thereby corresponds to the Hilbert-transformation

$$\mathcal{H}\{S(z)\} = \frac{1}{\pi} \int_{-\infty}^{+\infty} \mathrm{d}z' \, \frac{S(z')}{z - z'}$$

of the complex signal $S(z)$.

The imaginary part of a complex function can thereby be retrieved by the Hilbert-transform if it can be assured that the detected signal is analytic. By doing so, one can obtain the complete complex description of a wavefield including the correct phase, although only the intensity of the interference can be detected.

The question remains, how it can be ensured that a signal is analytic? There are several imaging modalities, which can ensure that their detected signal corresponds to an analytic signal and therefore they can obtain the whole information about the detected complex wavefield such as holography [75–77] and digital holography [78]. One of them is optical coherence tomography, which is explained in the following section.

2.4 Optical Coherence Tomography

Optical coherence tomography (OCT) was first introduced by Huang et al. in 1991 [79]. The basic idea of OCT is to use interferometric measurements of scattering samples to detect the depth of the scattering structures. Light with short coherence length is used, which is split to propagate a reference arm and a sample arm. The light of the reference arm is reflected back by a mirror after a known path length, while the light of the sample wave is backscattered from the sample. The two beams are then superimposed on a detector. If the difference between the optical path lengths covered by the two partial beams is less than the coherence length of the light, interference fringes can be detected. Depending on the depth of the sample structure, constructive, destructive, or something in between those two can occur.

In the sample, the object light is scattered back from positions where the refractive index is changing. Thus, variation of the refractive index make layered structures such as the retina visible for OCT.

There are in general two different approaches for generating the depth profile of the sample by the interference of the two partial beams. The historically first method is time-domain OCT (TD-OCT). Here the depth profile is recorded by a successive displacement of the reference mirror over time. The axial resolution of TD-OCT depends on the coherence length and increases , when the coherence length of the light source decreases (Fig. 2.1 a).

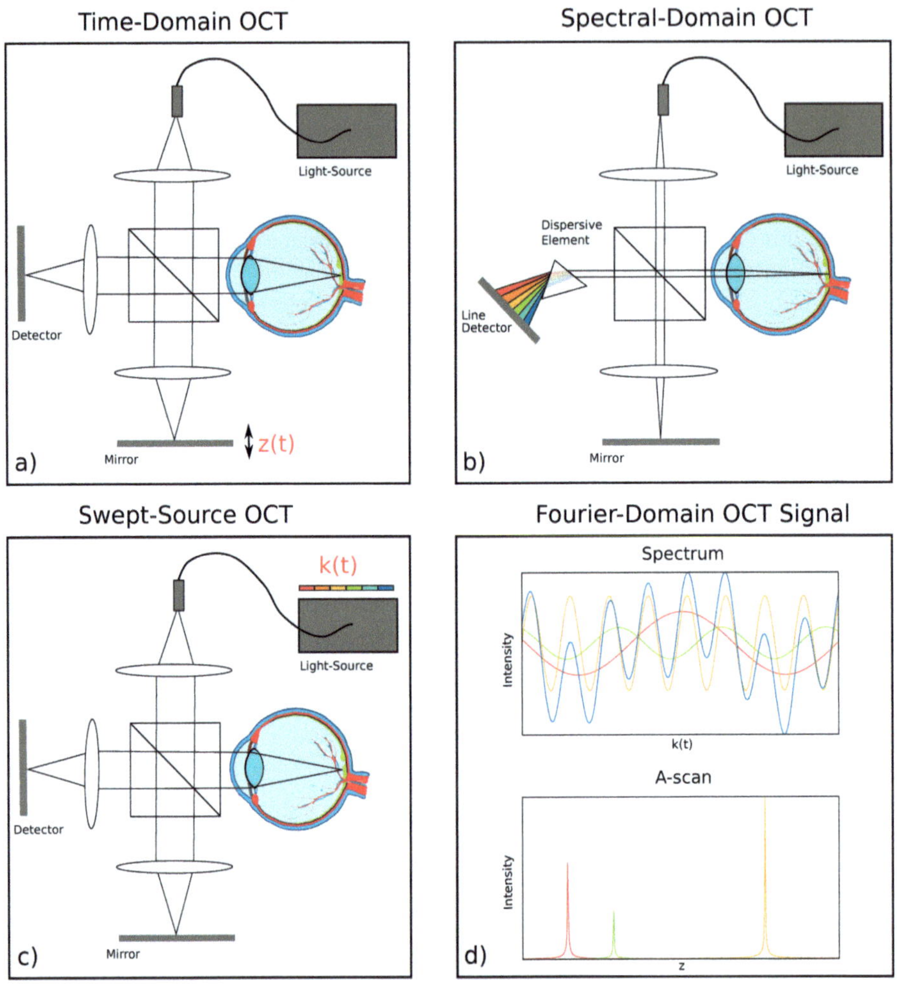

Figure 2.1: Schematic setup of different OCT systems. a) Time-domain OCT b) Spectral-domain OCT c) Swept-source OCT d) Spectrum detected by Fourier-domain OCT with the corresponding A-scan. Red, green, and yellow curves are the spectrum for the correspondingly colored single peak in the A-scan. The blue curve is the spectrum of all peaks

The other OCT approach is the Fourier-domain OCT (FD-OCT). Here, as the name suggests, the frequency spectrum of the depth profile is recorded by detecting the interference signal between the sample and the reference arm in a frequency-dependent way. This can be done in two different ways. In the spectral-domain OCT (SD-OCT), the interfering light is split into its spectral components by means of a wavelength selective optical element, such as a grating or a prism, and recorded by means of a line detector [80] (Fig. 2.1 b).

The swept-source OCT (SS-OCT), tunes the wavelengths of a broadband light source and thus records the interference signal over time [81] (Fig. 2.1 c).

If the sample light is backscattered at a certain depth, the different wavelength all have a different phase difference to the reference wave according to their frequency. This results in periodic constructive and destructive interference for a certain depth at the detector when tuning the wavelengths (Fig. 2.1 d). The greater the offset between the sample arm and the reference arm, the higher the frequency with which these two states change because the phase differences for different wavelength increases faster and faster for larger distances. Thus, the optical spectrum contains the frequency spectrum of the depth profile. Therefore, by Fourier transforming the optical spectrum, the depth information of the sample is obtained.

In the following section, the signal generation in FD-OCT is explained in more detail. This will be done analogously to the Ph.D. Thesis of Dierck Hillmann [82].

2.4.1 The OCT-Signal

As already mentioned in section 2.2, the the intensity of the interfering reference and sample wave is given by

$$I(x,y,k) \propto |R(x,y,k)|^2 + |O(x,y,k)|^2 + 2\mathfrak{Re}\{(R^*O)(x,y,k)\}, \tag{2.12}$$

according to Equations (2.4) and (2.5). Here R is the reference wave, O is the object or sample wave, x and y describe the lateral position of the wavefield, and k is the wavenumber. Measuring the intensity includes averaging over the high-frequency oscillations of the electromagnetic field. This temporal dependency of the intensity can be neglected. The reference wave can be further described by:

$$R(k) = A_R A_C(k) e^{\mathrm{i}kz_0}. \tag{2.13}$$

Here, A_R is a constant factor for the transmission in the reference arm. A_C is a wavenumber dependent factor for the amplitude and describes the shape of the spectrum, and z_0 is the total optical path length of the reference arm. The object wave can be described by

$$O(k) = A_O A_C(k) \int_{-\infty}^{\infty} \mathrm{d}z' \, \eta(z') e^{\mathrm{i}k(z_0+2z')}. \tag{2.14}$$

Here, A_O is the transmission in the object arm, z' is the axial position relative to z_0 and $\eta(z')$ is the depth depending scattering coefficient of the sample. The factor 2 is due to the fact that the path difference z' is passed through twice by the light; once on the way to the object and once on the way back. By inserting Equation (2.13) and (2.14) in Equation (2.12) the following results for the detected intensity:

$$\begin{aligned} I(k) =& S(k) A_R^2 \\ &+ S(k) A_O^2 \iint_{-\infty}^{+\infty} \mathrm{d}z \, \mathrm{d}z' \, \eta(z) \eta(z') e^{\mathrm{i}k2(z-z')} \\ &+ S(k) 2\mathfrak{Re}\left(A_O A_R \int_{-\infty}^{+\infty} \mathrm{d}z' \, \eta(z') e^{\mathrm{i}k2z'} \right) \end{aligned}$$

Here, $S(k)$ is equal to $A_C^2(k)$ and corresponds to the spectral intensity of the used light source. The first term or DC-term describes the interference of the reference wave with itself. The second term or the autocorrelation term describes the interference of object waves backscattered from different depths in the sample. Finally, the third or cross-correlation term describes the interference between the reference and the object wave, and thereby contains the actual information of the sample's depth-profile. An inverse Fourier-transform of the cross-correlation term is then used to reconstruct the depth-profile according to:

$$\begin{aligned} &\mathcal{F}^{-1}\left[S(k) A_R A_O 2\mathfrak{Re}\left(\int_{-\infty}^{+\infty} \mathrm{d}z \, \eta(z') e^{ik2z'} \right) \right] \\ =& \mathcal{F}^{-1}\left[S(k) A_R A_O \left(\int_{-\infty}^{+\infty} \mathrm{d}z' \, \eta(z') e^{-\mathrm{i}k2z'} + \int_{-\infty}^{+\infty} \mathrm{d}z' \, \eta(z') e^{\mathrm{i}k2z'} \right) \right] \\ =& \tilde{S}(z') A_R A_O * \left(\eta\left(\frac{z'}{2}\right) + \eta\left(-\frac{z'}{2}\right) \right) \end{aligned}$$

In the results, the depth-dependent course of the scattering coefficient and its complex conjugated counterpart is folded with the Fourier-transformed spectrum is obtained. Thus the Fourier-transformed spectrum can be considered as the axial point spread function (PSF). From this equation it becomes obvious, as already shown in section 2.3, that by Fourier-transforming the real part of a complex signal, not only the wanted depth-profile $\eta\left(\frac{z'}{2}\right)$ but also the complex conjugated mirror image $\eta\left(-\frac{z'}{2}\right)$ of the depth-profile is obtained. In this case, the complex conjugated counterpart is introduced to the signal, because only the modulus of the optical path length differences $|z'|$ can be detected. As long as it is not known which wave field had traveled through a longer optical path length, a positive difference z' is equally possible as a negative difference $-z'$. If, however, it can be ensured that one wave field has always a shorter optical path length than the other wavefield, the detected optical path length difference would be unique. Thus, if the optical path length in the reference arm is always chosen a little bit shorter than all optical path length in the object arm, then there will be only positive optical path length differences between the two arms. Thus, an analytic signal is generated and the whole information of the complex wavefield can be retrieved by the Hilbert-transformation.

2.4.2 Speckle Formation as a Hindrance for Phase Evaluation

Although by using OCT it is possible to detect the phase of the backscattered light in most cases it is still not used for a further interpretation of the sample. One main reason for that is the formation of speckle-noise in coherent imaging, which renders the interpretation of the phases difficult. To still use the phase for further evaluation, it is necessary to understand how the speckle arises. To this end, the effect of backscattering by structures in a volume below the resolution of the OCT system has to be understood.

For this reason, a completely plane structure is considered, which is perpendicular to the z-direction. Then all beams which are backscattered from the plane, would exhibit the same phase according to its axial position and the superposition of one voxel, would lead to completely constructive interference of all beams (Fig. 2.2 a). Thus the detected phase would yield information of the axial position of the surface with a sub-resolution precision.

However, most samples do not consist of perfectly even surfaces. Biological tissues contain small randomly distributed scatterers within a sub-resolution volume. The phase differences of the different scattered waves are randomly distributed as well and therefore the superposition of all waves follows a statistical process, which can be described by a random walk in the complex plane (Fig. 2.2 b). The resulting probability density of the intensity of the wavefield is thereby distributed according to an exponential function, while the phase is evenly distributed between $-\pi$ and π [72],

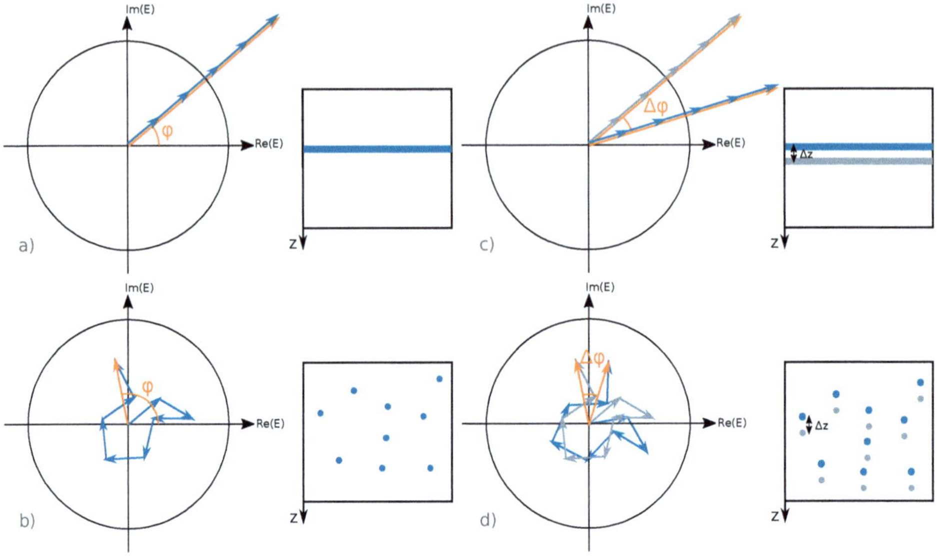

Figure 2.2: Superposition of multiple beams for a) an even surface perpendicular to the z-axis. All beams backscattered have the same phase and add up constructively. The superposition maintains the same phase; b) randomly distributed scatterers. The beams backscattered from the different scatterers have random phases, which add up according to a random walk in the complex plane, the resulting phase is random and the magnitude is small. c) and d) shows the influence of a systematic change of the scatterers for a z-plane and for randomly distributed scatters, respectively. A temporal referencing removes the random phase of the speckle, so that the phase difference corresponds to the systematic change of the sample.

which results in the formation of speckle noise. Thus, if the sample exhibits speckle noise, then the phase is randomly distributed and does not contain usable information on the sub-resolution structure of the observed sample.

Since the phase distribution in the scattered light is fixed to the tissue structure, phase and intensity will not change as long as the tissue is static. However, if the tissue structure is moving in z-direction the phase of the speckle pattern will change as well. As long as the random phase within a voxel stays the same meaning the single scatterers have the same random distances to each other, a temporal referencing of the phase will remove it. The phase change $\Delta\Phi$ will therefore contain information of the collective motion of all scatterers according to:

$$\Delta\Phi = 2k_0 n \Delta z. \tag{2.15}$$

By extracting the phase differences in a temporally changing speckle pattern, dynamic changes of the sample in axial direction, below the axial resolution can be detected (Fig. 2.2 c and d).

However, to ensure that the random phase of the speckle pattern is truly removed, it needs to be ensured that only the exact same speckle are referenced to itself over time. Bulk motion of the sample can, however, result in comparing different parts of the same speckle pattern or create gaps in scanned volumes.

To avoid or to correct the influence of such motion during the measurement one can evaluate only fast tissue changes or use a 3D imaging of the sample, which enables us to remove tissue motion afterward with a suitable co-registration of the data.

2.4.3 Full-Field Swept-Source OCT

Unlike the scanning OCT systems, full-field OCT (FF-OCT) does not use a focused beam that is scanned laterally across the sample for imaging, but a parallel beam that simultaneously illuminates all lateral position and records the scattered light using an area camera (Fig. 2.3). Since all lateral positions are thus recorded at once, sample movements during the acquisition of a volume cannot cause lateral gaps in the data, and sample movements between consecutive volumes can be referenced with the necessary accuracy through co-registration. Both full-field time-domain OCT (FF-TD-OCT) and full-field swept-source OCT (FF-SS-OCT) have been established for

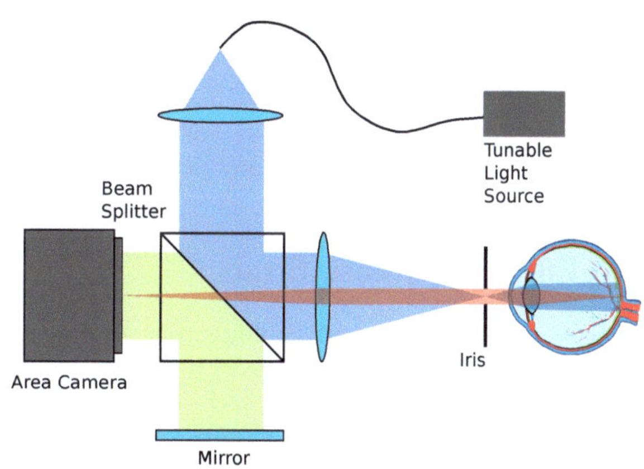

Figure 2.3: FF-SS-OCT setup: The light of a tunable light source is collimated and split into a sample beam (blue) which illuminates the retina, and a reference beam (green). The light of the reference beam is reflected by a mirror onto a area camera where it is superimposed with the light back scattered from the sample (red).

certain applications. FF-TD-OCT is usually used for imaging with very high lateral resolution [83–86]. Spatially incoherent light is used in combination with a Linnik interferometer. Thus, the interference of multiply scattered photons is suppressed. By using a broadband, highly incoherent light source, an axial resolution below one micrometer can be achieved. This technique has a wide range of applications in ex-vivo imaging, where its high resolution approaches that of classical histological sectional images. Besides, FF-TD-OCT provides the possibility to add a functional contrast of biological samples to the image information by spectral analysis of temporal scatterer fluctuations [87, 88]. The disadvantage of this imaging technique is that the depth still has to be scanned mechanically and that the photon yield is relatively poor due to the axial and lateral gating, which means that relatively long imaging times are required. Thus, although parallel illumination provides lateral phase stability, axial phase stability is difficult to achieve, at least for in-vivo measurements.

By recording the full spectrum instead of the depth profile itself, FF-FD-OCT allows parallel axial acquisition in addition to the lateral parallelization. Since the flat illumination of the FF-OCT means that no spatial dimension is available for the detection of the depth spectrum, FF-FD-OCT is implemented as a swept-source version.

This technique was first demonstrated by Považay et al. in 2006 [89]. Due to the parallel recording in all three spatial dimensions very high acquisition rates can be reached. The limiting factor for the acquisition speed is here, unlike in scanning OCT systems, not the sweep rate of the light source but the frame rate with which the area camera detects the depth spectrum. Current high-speed area cameras allow, with a sufficiently large field of view and measurement depth, volume rates of more than 100 Hz, which is sufficient to record in-vivo retinal volumes completely phase-locked. This was first demonstrated by Bonin et al. in 2010 [90]. Since then the detection of phase-stable OCT data has proven to be useful in various areas. A major advantage of phase-sensitive imaging is, for example, the numerical correction of imaging errors.

2.4.4 Cause and Effects of Phase Errors

Despite the fast detection of phase-locked volumes, phase errors still occur, which decrease the image quality of the data. Two types of phase errors can be distinguished in FF-SS-OCT. Firstly, axial movements during recording cause a phase chirp, which is equivalent to dispersion differences in sample and reference arm. Such phase errors cause axial smearing of the data. Secondly, there are phase errors caused by imprecise optics of the setup or especially of the eye. This results in aberrations which lead to lateral smearing of the data.

In an ideal setup, the interference spectrum consists of an evenly change of constructive and destructive interference, meaning a specific frequency for a single depth. Thus, for a thick tissue the interference is given by

$$I(k) = 2\gamma S(k) A_R A_O \int_{-\infty}^{+\infty} \mathrm{d}z\, \eta(z) e^{-\mathrm{i}2kz}.$$

The spectra, however, can be disturbed if the optical path length z is not independent of the wavenumber. Thereby, the optical path length can be changed by two factors; the physical distance l the light travels and the refractive index n of the medium it penetrates.

While for a static sample the physical length is independent of the wavenumber, the refractive index is typically not. This means, that different wavelengths travel at different speeds through the media. This phenomenon is known as dispersion. If there is a refractive index mismatch between the reference and the sample arm, then the measured optical path length differences will change for different wavenumbers. Similarly, if a non-static sample is considered, axial motion during the wavenumber sweep leads to the same effect. Thereby the original spectra is changed according to:

$$I_{\mathrm{disp}}(k) = 2\gamma S(k) A_R A_O \int_{-\infty}^{+\infty} \mathrm{d}z\, \eta(z) e^{-\mathrm{i}2k(z+z_{\mathrm{disp}}(k))} \tag{2.16}$$

$$= 2\gamma S(k) A_R A_O e^{-i\Phi(k)} \int_{-\infty}^{+\infty} \mathrm{d}z\, \eta(z) e^{-\mathrm{i}2kz}. \tag{2.17}$$

Therefore both, dispersion and axial motion during the wavenumber sweep introduce an additional phase factor or phase error $\Phi(k) = 2kz_{\mathrm{disp}}(k)$. Thus $H_{\mathrm{disp}}(k) = e^{-i\Phi(k)}$ resembles the corresponding complex phase error function. Depending on the order of this phase error, different effects on the image occur. A constant phase error does not affect the image amplitudes itself. If the phase error is of first order in a polynomial approximation, this only results in a wrong frequency in the spectra, and thereby the image is shifted to another depth, without a change in the image (Fig. 2.4 b). However, if the phase error is of a second or higher-order the frequency spectrum gets chirped and the corresponding image is axially blurred (Fig. 2.4 c).

Similar effects to these axial phase errors have lateral phase errors or aberrations. As mentioned in section 2.2 lenses perform a Fourier-transform of the wavefield, which is located in the negative focal plane or aperture plane. A single point on the optical axis in the object plane thus corresponds to a planar wavefront in the aperture plane (Fig. 2.5 b). This wavefront can be deformed by imprecise optics of the setup or the eye so that the image spectrum is superimposed by a two-dimensional phase error function. The aberrated wave field in the aperture plane $\tilde{U}(\boldsymbol{k})$ can thereby be expressed as

$$\tilde{U}(\boldsymbol{k}) = H_{\mathrm{aber}}(\boldsymbol{k}) \tilde{U}_0(\boldsymbol{k}), \tag{2.18}$$

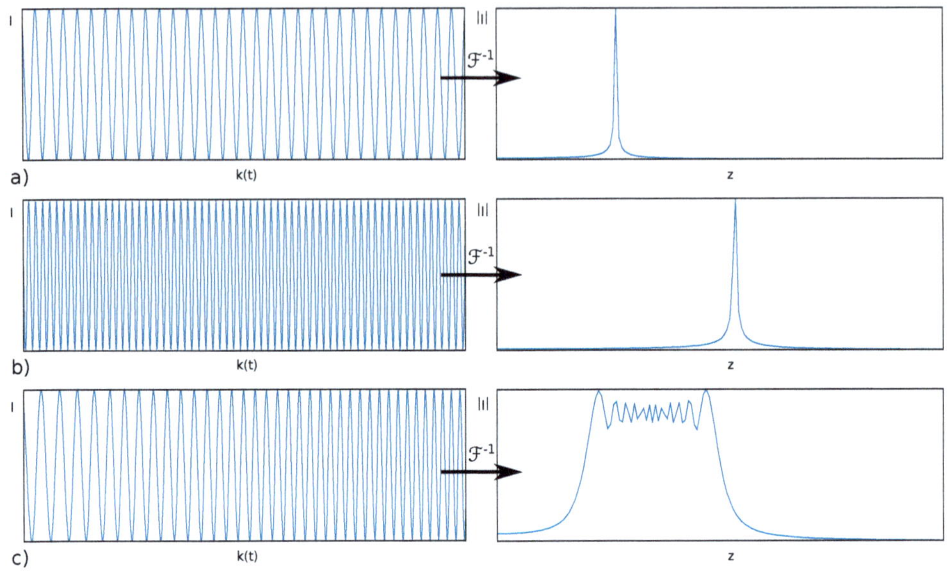

Figure 2.4: Influence of an axial phase error on OCT data: a) A single frequency in the spectrum correspond to a single axial peak in the depth profile. b) Linear phase error results in a different frequency. This only moves the structure in depth. c) quadratic phase error result in a chirped frequency. The corresponding depth profile is axially smeared, since the signal seems to come from different depth for different parts of the spectrum.

with $H_{\mathrm{aber}}(\boldsymbol{k}) = \exp(-\mathrm{i}\Phi(\boldsymbol{k}))$ being the amplitude transfer function or the lateral phase error, $\tilde{U}_0(\boldsymbol{k})$ being the original, non-corrupted wave field, and $\boldsymbol{k} = (k_x, k_y)$.

Just as with axial phase errors, an additional phase error function on the lateral frequency spectrum can influence the image quality. Accordingly, a first-order phase error function leads to a lateral shift of the image (Fig. 2.5 c) and higher-order phase errors lead to lateral smearing of the point spread function (PSF) (Fig. 2.5 d).

From Equation (2.17) and (2.18) it become obvious, that once the phase error is known, it can easily be removed in coherent imaging by multiplying the spectra with the complex conjugate of the phase error term $H_{\mathrm{disp}}(k)^*$ or $H_{\mathrm{aber}}(\boldsymbol{k})^*$, i.e., the inverse phase error function. Thus, in order to correct for those phase errors one only needs a strategy to detect the present phase error.

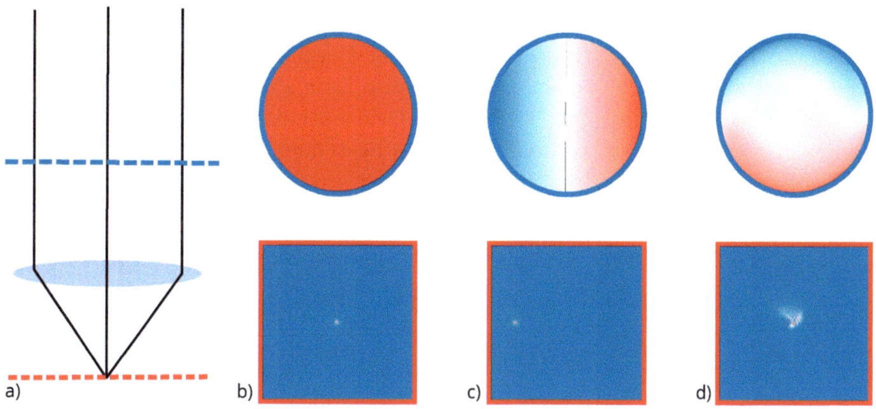

Figure 2.5: Influence of lateral phase errors. a) Optic setup with image plane (dashed red) and corresponding Fourier plane (dashed blue). b) Phase in the aperture plane (upper graph) is constant and the corresponding PSF in the image plane is narrow (lower graph). c) A linear phase error function in the aperture plane laterally shifts the PSF in the image plane. d) A higher order phase error function in the aperture plane blurs the PSF in the corresponding image plane.

To a certain extent, there are hardware-based possibilities to compensate for these phase errors. For example, by the proper choice of optical elements in the setup, dispersion mismatches in the reference and sample arm can be minimized. However, this cannot correct for axial phase errors caused by sample motion.

In contrast, aberrations can be corrected by adaptive optics. Such setups measure the shape of the wavefront with a sensor and compensate for the deformation by adapting the surface of a deformable mirror within the beam path. Since aberrations can change extremely fast due to eye movements, fast control loops are necessary. This makes hardware-based compensation of phase errors expensive, difficult to adjust and they cannot correct phase errors of arbitrary order.

One of the great advantages of phase-locked imaging is that such phase errors no longer have to be corrected with expensive hardware. Since the phase-sensitive acquisition gives access to both the complex image data and their complex spectra, the phase errors can be corrected numerically after the acquisition.

The main task for the correction is to determine the phase error function. This can be done directly from the acquired data by different approaches. For example, the so-called phase gradient auto-focus algorithm, adapted from synthetic aperture radar, which works with statistical estimation of the phase error by using a dominant structure in the sample as some kind of guide-star, has been shown to correct for axial phase errors [66]. Methods using a guide-stare for lateral correction of phase errors have also been demonstrated [91, 92]. Another deterministic approach obtains the phase error via cross-correlation of sub-spectra. This has also been shown for both axial [67] and lateral phase error functions [69, 93, 94]. Also, an optimization, that uses the image sharpness as a criterion can be used, which has been done both for correction of axial as well as for lateral phase errors [68, 95].

2.5 The Nervous System

In addition to the technical basics for functional imaging, for this thesis, it is equally important to have a sufficient understanding of the general functional principals of the nervous system, the structure of the retina, its neural circuity, and the signal processing that takes place in the individual cells to ensure an interpretation of the functional data, which are recorded by FF-SS-OCT. Therefore, the following section will start with a short overview of the general working principles of the nervous system, followed by an overview of the retina, as well as of some selected signaling pathways and information processing steps in the individual cell types.

The nervous system is responsible for the information processing of the organisms. It encodes, processes, and stores information from the environment and uses it to adapt the physiology and behavior of the organism to this information. The nervous system in most vertebrates can be divided into two parts: the peripheral and the central nervous system (PNS and CNS, respectively). In Humans, the CNS consists of the brain and the spinal cord, whereas the PNS is responsible for communication between the CNS and the rest of the body.

The neurons form the functional units of the nervous system. These are specialized cells that can process information and then transmit this information to other cells by changing the electrical potential of their plasma membrane. Neurons can look very different according to their task. However, they can always be traced back to a few basic structures and functional principles.

2.5.1 Membrane Potentials

All neurons process information and pass it on across their plasma membrane by changing the electrical potential. At rest, most neurons have a potential of about $-60\,\mathrm{mV}$. Small changes in the resting potential are very sensitively perceived by the neurons and can ultimately result in the generation of an action potential. The action potential is a short intensive change of the resting potential and is used by neurons to transmit information, without loss, over long distances to other neurons. However, some neurons, especially receptor cells, transmit their information not with action potentials but with so-called generator potentials. Both signals have their advantages and disadvantages. To illustrate this, the different potentials are briefly explained.

The plasma membrane of the cells is impermeable to ions inside and outside the cell. However, it contains a large number of ion channels and ion-pumps that can transport the ions across the cell membrane. The sodium-potassium-exchanger is one of the key membrane proteins for the resting potential. It actively transports sodium ions (Na^+) against their concentration gradient out of the cell and exchanges them for potassium ions (K^+) from the external environment using chemical energy. This results in a high K^+ concentration inside the cell.

The passive transport of potassium is governed by ion channels. Ion channels are pores formed by channel proteins that allow ions to diffuse across the cell membrane. In general, they are selective, i.e., only a specific type of ion can diffuse through a specific channel type. The direction and strength of the ion flux depend on the concentration gradient and the electrical voltage across the cell membrane. In a non-exited state, the potassium channels are open much more often than the other ion channels. Since the K^+ concentration inside the cell is much higher than outside due to the sodium-potassium-exchanger, K^+ leaks out of the cell and leave uncompensated negative charges in the cell. The K^+ efflux thus builds up an electrical potential across the cell membrane, which draws K^+ back into the cell. The electrical potential at which the electrical force and the force by the concentration gradient are balanced, without a net flux of K^+ across the membrane, is the potassium equilibrium potential. This approximately corresponds to the resting potential. In reality, the resting potential is a little more positive compared to the potassium equilibrium potential, since not only K^+ is involved in its build-up. This resting potential can now be changed by external influences. Channel proteins are primarily responsible for the change in the membrane potential.

A large part of the channel proteins located in the cell membrane are so-called gated channels. They only open under certain conditions, for example at certain voltages or when they have bound ligands. Depending on the ion for which these gated channels are selective, the membrane potential becomes more negative (hyperpolarization) or more positive (depolarization) when they open. This allows the neuron to change its membrane potential in response to external stimuli.

If the membrane potential depolarizes by only a few mV to about $-55\,\mathrm{mV}$ then voltage-gated sodium channels open. Both the concentration gradient and the electrical potential attract Na^+ into the cell interior. As a result, the cell depolarizes to about $+50\,\mathrm{mV}$ within a very short time. However, the voltage-gated sodium channels are only open for a very short time, and shortly after their opening, voltage-gated potassium channels open, which leads to a regeneration of the resting potential. Therefore, the depolarization only lasts for about $2\,\mathrm{ms}$ before the resting potential is restored. The resulting short spike is the action potential. The organized interaction of the voltage-dependent channels ensures that the action potential always has the same amplitude and duration. It also ensures that it is self-generating along the plasma membrane, as it depolarizes the surrounding membrane areas and thus initiates a new action potential. Thus, information can be carried along the plasma membrane of a cell over long distances without losses.

The action potential consists effectively of only two states and the only parameter modulating the information transfer is the frequency, with which the action potentials are generated.

In addition to the action potentials, there are the so-called generator potentials. They are generated in many receptor cells, i.e., cells that react sensitively to sensory stimuli outside the body. They are generated by opening or closing unspecific cation channels and provide a potential change, whose amplitude can be scaled with the intensity of the stimulus. They can also be of very different duration varying between 40 ms and 4 000 ms. Thereby the potential changes are not binary, but can contain analog information. However, since the potentials cannot generate themselves, it cannot run along the neuron membrane without loss. Therefore, it can only be used to pass on information over short distances.

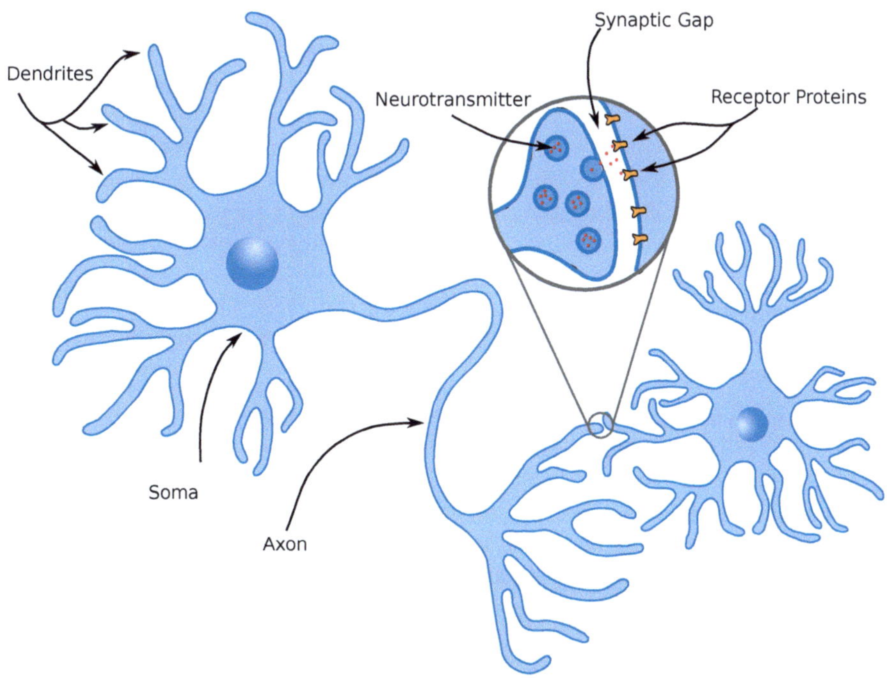

Figure 2.6: Schematic drawing of two neurons: The circular close up shows the synaptic connection of the two neurons, where the synaptic connection is a chemical synapse.

2.5.2 Structure of Neurons

As with functionality, there are a few basic structures that are common to almost all neurons. A typical structure of two neurons and their connection is displayed in Figure 2.6. The cell body (soma) of neurons contains the nucleus and most organelles for the cells. Several branch-like outgrowths originate from the soma. Most of these outgrowths are so-called dendrites, which branch out towards other neurons. They receive information from these and transmit it to the cell body by means of electrical potential changes along their membrane. An exception to this are receptor cells because they do not receive information from other cells but detect sensory stimuli such as photons. They do not have dendrites but specialized units that detect the sensory stimuli and convert them into a membrane potential change.

In the soma, the incoming information of all dendrites is processed, so that a corresponding signal can be passed on to other neurons. Therefore only one output signal is generated for each neuron. This is done at the soma membrane close to the axon. The axon is also a branch-like outgrowth at the soma of the cell. It is usually longer than the dendrites. The output signal of the neuron is transferred via the axon. At the end of the axon, it can branch out to form contacts to different cells. The contact points of the axon with the following neuron are called synapses. In general, there are two different types of synapses; electrical and chemical. In most vertebrates, mainly chemical synapses are formed. Chemical synapses are not directly connected to the postsynaptic neurons. Between the plasma membranes of these two cells, there is a gap of about $20 - 40\,\text{nm}$; the synaptic gap. The chemical synapses are filled with many vesicles that contain a high number of neurotransmitters. When an electrical signal reaches a chemical synapse, the vesicles fuse with the plasma membrane and release the neurotransmitters into the synaptic gap. There, the neurotransmitters then diffuse to the postsynaptic membrane where they bind to corresponding receptors.

How the postsynaptic cell responds to the neurotransmitter depends on the type of the receptors. Depending on whether they can be blocked or activated by the neurotransmitters and for which ions they change the permeability across the membrane, the neurotransmitter cause hyperpolarization or depolarization, thus reducing or increasing the probability of triggering an action potential in the postsynaptic cell, respectively. In this way, synapses can be distinguished into inhibitory and excitatory synapses.

Furthermore, the receptor proteins can be divided into two major classes; ionotropic and metabotropic receptors. Ionotropic receptors are ion channels themselves - a binding of a neurotransmitter thus leads to an immediate ion current. Thus, they enable a fast and short-lived signal. Metabotropic receptors, on the other hand, induce a signal cascade in the postsynaptic cell, which causes a change in the ion channels. The resulting signal is therefore slower and usually more long-lived. [96]

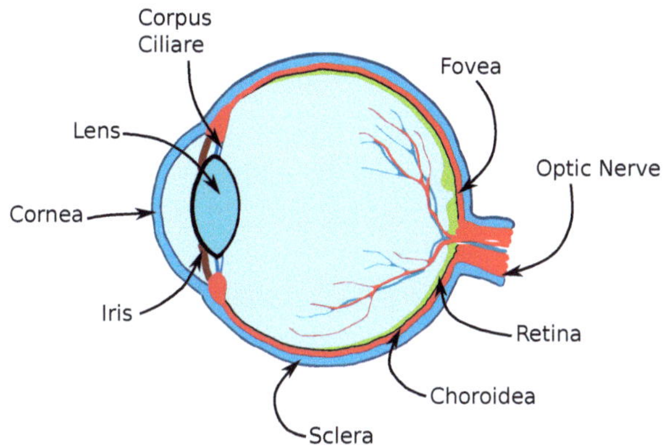

Figure 2.7: Schematic illustration of the anatomy of the eye.

2.6 Anatomy of the Eye

Our eyeball is about 2 cm in diameter. A schematic side view of it is shown in Fig 2.7. It has a layered structure. The outer layer is built by the sclera and the cornea. The sclera is a white structure, which is enriched with collagen fiber bundles. It thereby ensures that the spherical structure of the eye is kept. In the front, the sclera merges into the cornea, which is a transparent structure. Since the transition from air to the cornea states the highest refractive index change in our visual system, the cornea is responsible for the major part (+43 dpt) of the optical refractive power. The second layer is a pigmented layer. It can be subdivided into three different structures: the choroidea, the corpus ciliare and the iris. The choroidea is situated at the dorsal side of the eyeball. It is responsible for the retinal supply with substance. The choroidea merges into the corpus ciliare in the front. This structure holds the lens. While contracting the musculus ciliare, the lens gets a more spherical form, which enables us to focus on structures near us by creating a sharp image of it onto the retina. The last part is the iris, which works as an aperture. Therefore it covers the whole lens except for a small spot in the center, which is known as the pupil. The size of the pupil can vary to regulate the amount of light, that reaches the back of the eye. For bright light conditions, the size is below 2 mm in diameter, while for dim light conditions the diameter can expand to about 8 mm. This pupil reflex can be blocked by eye drops containing atropine, to receive a temporally enlarged pupil diameter independent of the surrounding light condition. The innermost layer of our eye (tunica interna bulbi) is the retina. The retina contains the light-sensitive neurons and some other higher-order neurons, which transmit the visual information via the optic nerve to the brain. The single components, the structure, and the function of the retina are explained in more detail in the following sections. [97]

2.6.1 The Retina

Unlike the other sensory system in the human body, the retina develops during embryogenesis from the neuroectoderm and therefore is a part of the CNS [96]. This is why the retina not only converts the light stimuli into a series of membrane potentials but is also responsible for the first post-processing steps of the visual information.

The retina has a strictly layered structure (see Fig. 2.8). The outer layer, which is closest to the choroidea, is the retinal pigment epithelium (RPE). The RPE is a single cell layer, where the individual cells are connected via tight junctions, thereby creating an impervious barrier between the choroidea and the neurons of the retina. It can therefore be considered as the blood-brain-barrier (BBB) of the eye. It is responsible for the supply of nutritive substance for the photoreceptor cells, as well as for the removal of their waste. Additionally, the RPE cells contain a huge amount of melanin molecules, that lead to strong scattering and absorption of the incident light, thereby minimizing further penetration of light in the underlying tissue.

The other cell layers above the photoreceptor cells, are supplied by a vessel system that reaches the retina via the optic nerve. Thereby, larger vessels are situated in the ganglion cell layer (GCL), which stratify in a more narrow capillary system in the plexiform layers of the retina.

The layer above the RPE is built by the photoreceptor cells. They are the neurons, that detect the incident photons and convert them into a change of membrane potential and finally into a chemical signal, that is transferred to the next neuron layer. The

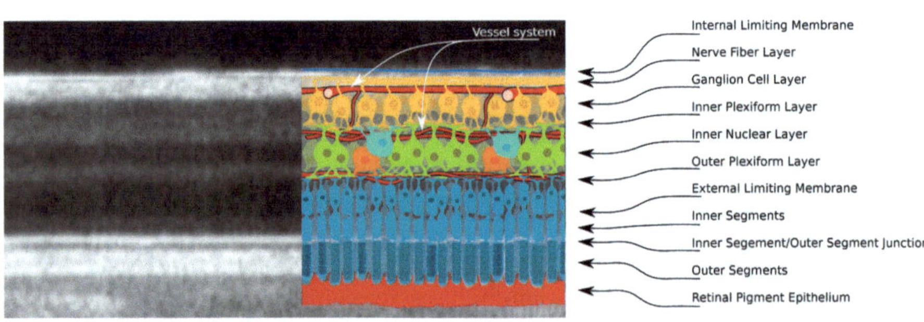

Figure 2.8: Schematic illustration of the retinal layer structure correlated with an FF-SS-OCT B-scan (averaged over 70 co-registered volumes) in 2 mm periphery.

photoreceptors can be subdivided into two different classes: the rods and the cones. The rods are very sensitive to dim light and can even detect single photons. They are responsible for our night vision. In contrast, the cones are responsible for our vision under daylight conditions. They are less sensitive than the rods, however, they react faster and feature different pigments, that have different spectral sensitivity across the visual spectrum. Thus they are also encoding for our color perception. Most mammalians have two types of cones, one that is sensitive to blue light and one, that is sensitive to green light. Primates have an additional pigment, that absorbs light preferable in the red range of the spectrum.

The outer plexiform layer (OPL) is situated above the photoreceptor cells and contains their synaptic connections with the secondary neurons the horizontal and bipolar cells. Their cell bodies as well as the ones from an additional neuron class, the amacrine cells compose the inner nuclear layer (INL). All of those cell classes can be further subdivided into sub-classes or types that are sensitive to specific features of the photoreceptor signal. The amacrine and bipolar cells build a synaptic connection with the third neuronal layer, the ganglion cell layer (GCL), in the inner plexiform layer (IPL). The ganglion cells can, again, be further sub-divided into different ganglion cell types, that are further specialized for features in the visual signal. They convert their input signal into a series of action potentials, which they transmit via 1-2 million axons to the visual cortex. The ganglion cell axons compose the last layer of the retina; the nerve fiber layer (NFL). All axons gather in the optic nerve, which penetrates the retina in the axial direction so that the visual information can be transmitted to the brain. Since at this particular spot no photoreceptors are present, we cannot see with this part of the retina. It is therefore also called blind spot. A systematic overview of the retinal layered structure in correspondence with the depth profile of an FF-SS-OCT volume is shown in Figure 2.8.

In nearly all mammalian eyes, the retina is dominated by the rod photoreceptors. For example, the primate retina exhibits about 120 million rods and only 7 million cones, although, primates preferable uses their retina in daylight conditions. Additionally, these two types are distributed very inhomogeneously in the retina. The cones have their highest density in the fovea, which is a rod-free area. The cone density decreases with the eccentricity, while the rod density increases (see Fig. 2.9). In the periphery, the bipolar and ganglion cells integrate the signals of many photoreceptors especially rods. This integration becomes smaller within the neighborhood of the fovea and reaches its minimum in the fovea, where there is a 1:1:1 connection of ganglion cells to bipolar cells to photoreceptors. This makes the fovea the spot of the highest acuity in the retina. Although this results in a high need for bipolar and ganglion cells, those two layers are missing in the fovea. The ganglion and bipolar cells corresponding to the cones in the fovea are shifted aside. Thereby the incidence light can reach the photoreceptor cells directly without traveling through the GCL and INL first.

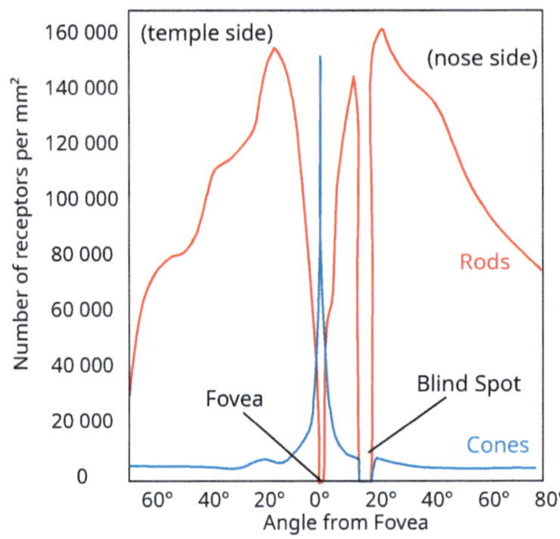

Figure 2.9: Distribution of rod and cone density over the retina. The figure was adapted from [98].

In the following sections the visual sensitivity and information processing is discussed in more detail. This starts with a short section about the spectral sensitivity of the retina followed by a detailed introduction of the different cell classes in the retina.

2.6.1.1 Photometric Units

Before discussing the function of the individual neurons in the retina in more detail, this section will first discuss photometric units that are important for a physiological interpretation of retinal stimulation.

Due to the absorption properties of the cones and rods, the eye is able to perceive wavelengths in the range of about 380 nm to 830 nm. This means that when light of this wavelength reaches the retina, the photoreceptors are stimulated. However, the retina is not homogeneously sensitive for all wavelengths in this range, but the sensitivity decreases towards the edges of the spectrum. Because of this, stimuli with the same irradiance but a different wavelength or spectrum are perceived as differently bright. In order to interpret a light stimulation on a physiological level, it is therefore common to convert the radiometric units of the stimulus into photometric units (luminous or luminance), which measures the perceived brightness.

The conversion is done by an empirically determined luminous efficiency function $V(\lambda)$, which describes the relative spectral sensitivity of the eye [99]. It is defined in the range from 390 nm to 830 nm and has its maximum of one at 545 nm, where the human eye is most sensitive. The conversion from the irradiance E_e $[\frac{\mathrm{W}}{\mathrm{m}^2}]$ to the illuminance E_v $[\frac{\mathrm{lm}}{\mathrm{m}^2}]$ is done by the following equation.

$$E_v = 683\frac{\text{lm}}{\text{W}}\int_0^{\infty} \mathrm{d}\lambda\, V(\lambda)E_e(\lambda) \tag{2.19}$$

Here $683\frac{\text{lm}}{\text{W}}$ is a conversion factor, which results from the historical definition of the lumen.

However, since it is difficult in most cases to determine irradiance or illuminance on the retina directly, the definition of luminance L_v $[\frac{\text{cd}}{\text{m}^2} = \frac{\text{lm}}{\text{sr}\times\text{m}^2}]$ is often used. The luminance is defined by the spatial and directional derivative of the light flux emitted by a light source. Unless there is scattering or absorption, the illuminance does not change during light propagation. Therefore, the illuminance at the retina is connected with the luminance at the cornea via the pupil area A_p and the distance from the pupil to the retina D:

$$L_v = \Phi_v \frac{D^2}{A_p} \tag{2.20}$$

At luminance levels of 10^{-6} cd/m^2 to 10^{-2} cd/m^2, vision is dominated by rods (scotopic range). A luminance of around 10^{-5} cd/m^2, for example, corresponds to the stimulation of the retina by an overcast night sky. At luminances in the range of 10^{-2} cd/m^2 to 10 cd/m^2 both rods and cones are active (mesopic range). An example for mesopic condition is the stimulation by a night sky at full moon, which corresponds to 0.1 cd/m^2. At luminances in the range of 10 cd/m^2 to 10^6 cd/m^2, photopic vision occurs, i.e., vision is mainly determined by the cones. The stimulation caused by a clear sky during the day corresponds approximately to a luminance of 8000 cd/m^2.

2.6.1.2 Photoreceptor Cells and Phototransduction

The photoreceptor cells are highly organized, elongated cells. The dorsal part of the photoreceptor cells, which is nearest to the RPE, is the outer segment (OS). The OS is densely packed with membrane discs. The inter disc spacing is reported to be between 28 nm and 32 nm for different species [100, 101]. This sub-unit of the photoreceptor cells is responsible for detecting the incident light and converting it into a generator potential. Rods and cones can easily be distinguished by the morphology of their OS. The rods OS are cylindrical with an average length of 25 µm and a diameter of 2 µm [102]. The cones have a conical shape with a diameter of 3 µm at the base and 1 µm at the tips. They are also shorter than the rod OS and have an average length of 13 µm [103]. While the discs of the rod OS are completely isolated from the plasma membrane and stored in the cylindrical shaped OS, the discs in the cone OS are caused by invagination of the plasma membrane, which is still connected to the disc membrane (Fig. 2.10). The OS of both, rods and cones, is connected to the inner segment via a narrow membrane tube called connecting cilium. The inner segments are enriched with mitochondria and therefore responsible for the power supply of the cell. The inner

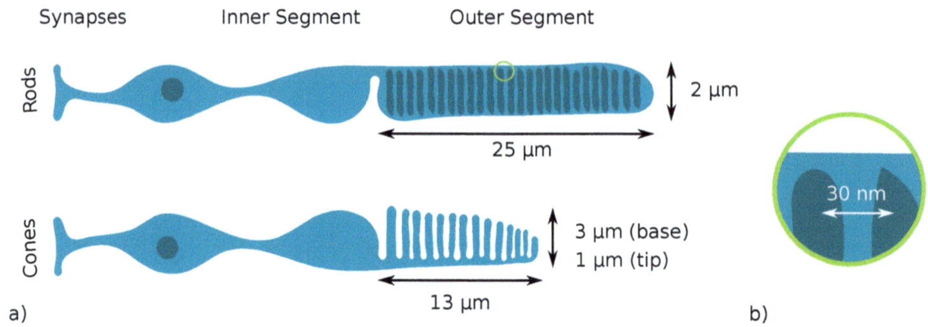

Figure 2.10: a) Schematic illustration of the structure of rods and cones. b) Closeup of the membrane disc in the rod outer segments. The information on the size of the outer segment and the inter disk distance was taken from [100, 102, 103].

segment merges into the nuclear layer where the nuclei of the photoreceptor cells are situated. After the nuclear layer, the photoreceptor cells built their synaptic connection to the subsequent bipolar and horizontal cells.

The photoreceptor cells are responsible for detecting the light and converting it into a chemical signal that, after several amplification steps, leads to a hyperpolarization of the photoreceptor plasma membrane, which can be considered as a generator signal. The membrane potential travels along the plasma membrane and leads to a suppression of the release of the neurotransmitter glutamate at the synaptic ending, which is detected by the second neuronal layer of the retina.

The main part of the phototransduction happens in the OS. The different steps of the phototransduction cascade are displayed in Fig 2.11. In the following, the rod phototransduction will be discussed in more detail. In general, the phototransduction cascade is the same for cones, however, it differs in the molecule concentration, the reaction rate of the single reactions, and the shutdown time of the different molecules, resulting in a faster and less sensitive response in cones.

The first activated molecule is a chromophore, that is enriched in the discs membrane with a membrane concentration of $25\,000\,\mu m^{-2}$, with around 1 000 discs for a mammalian rod OS. This equates to a concentration of $\sim 3\,mM$ in the OS [104]. This chromophore is built by a protein called rhodopsin, which binds a retinal molecule in its cis-configuration. An isomerization of retinal after light absorption leads to a conformation change of the rhodopsin, which in turn results in a separation of the rhodopsin molecule and the retinal. The separated rhodopsin is now activated (R^*). In the enzymatically activated state R^* binds and activates the guanosine triphosphate (GTP) binding protein, transducin (T), with an activation rate of $240\,s^{-1}$ [105, 106].

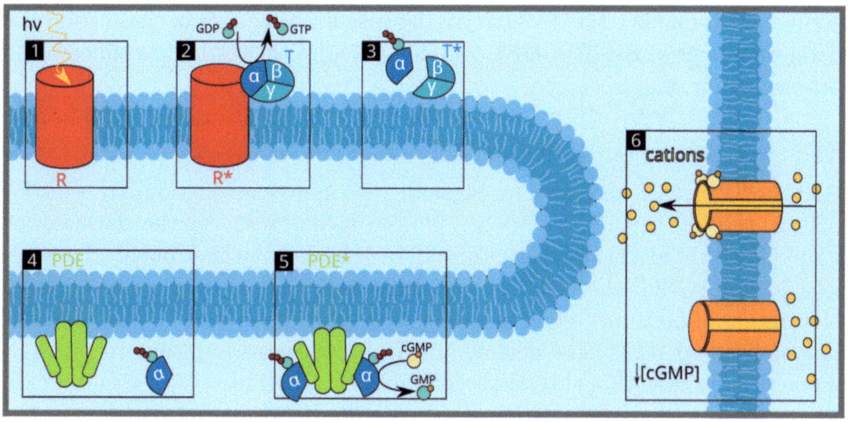

Figure 2.11: Systematic overview of the phototransduction cascade in the photoreceptor OS. 1) Inactive rhodopsin molecule (R), which contains the cis-retinal. 2) Retinal isomerates to all-trans retinal after absorption of a photon and thereby activates the rhodopsin molecule (R*). In its active states it binds transducin (T) and exchanges its GDP with GTP. 3) The alpha-unit of the transducin molecule is thereby activated (T*) and dissociate from the rest of the molecule. 4) The activated transducin subunit binds and activates a PDE molecule. 5) The activated PDE hydrolyzes the free cGMP to GMP 6) Due to the decline in cGMP concentration the CNG-channels in the plasma membrane closes, which inhibits the photocurrent.

During the lifetime of R* (80 ms), about 20 T molecules can be activated by a single R* molecule [107]. This is the first of three amplification step in the phototransduction cascade. The proportion of R and T in the cells is about 10:1 [108, 109].

In the darkness, T is composed of three different proteins one of which binds guanosine diphosphate (GDP). While T is bound to R*, GDP is exchanged for GTP. Once this is done, the activated subunit (T*) dissociates from the other two proteins and binds the enzyme phosphodiesterase (PDE), which is also anchored in the disc membrane [110]. The bond between T* and the PDE subunit activates the enzymatic PDE* that catalyzes the hydrolyzation of cyclic guanosine monophosphate (cGMP) to guanosine monophosphate (GMP). PDE is only activated as long as it is connected to T*. Therefore at this point in the phototransduction cascade, there is no amplification. The density of PDE is about 1-2% of the rhodopsin concentration [101, 108, 109]. Therefore, the phototransduction should be in saturation when only 0.1-0.05% of the rhodopsin molecules are activated. The T* lifetime is at least 2.5 times longer (200 ms) than the

R* lifetime [107], making the T* activity the rate-limiting step in phototransduction. The catalytic power of the T*-PDE* complex provides the second amplification of the phototransduction cascade.

PDE* hydrolyzes cGMP with a rate that is close to the limit set by aqueous diffusion [105, 111]. Thereby the intracellular cGMP concentration drops rapidly.

cGMP mediates as a second messenger in the phototransduction, transferring the signal from the disc membrane to the plasma membrane. In the darkness cGMP binds to cGMP gated cation channels and thereby keeps them open. These cyclic-nucleotide-gated (CNG) channels are embedded in the rod OS plasma membrane with a concentration of $500\,\mu m^{-2}$ corresponding to a total number of 6×10^4 molecules in the mammalian rod [112]. The decline in cGMP concentration leads to rapid closure of the open channels with a sub-millisecond delay [113].

When they are open, the CNG channels carry a steady inward current of cations (mainly sodium 80%, calcium 15%, and magnesium 5%) [114], which is driven by an electrochemical gradient. The electrochemical gradient in turn is caused by a continuous Na^+ efflux by a sodium-pump in the inner segment, which continuously transfers Na^+ against their concentration gradient from the cell inside into the external environment. This continuous electrical current in the outer segment in darkness is called dark current [115].

Once light initiates the phototransduction cascade, the CNG-channels close, which prevents the cations from entering the OS (third amplification), while there is still an efflux from cations via sodium pumps into the inner segment. This results in a hyperpolarization of the plasma membrane also know as the photocurrent. This behavior is rather atypical for neurons, which are commonly activated by a depolarized membrane potential. In fact, the photoreceptor themselves can be considered as activated in the dark states, since they release the neurotransmitter glutamate to the connected neurons during darkness while the hyperpolarization of the plasma membrane suppresses this release. The huge amplification of the phototransduction ensures that in rods single photoisomerization leads to a detectable photocurrent. The small lifetimes of the molecules participating in the signal cascade combined, with a fast recovery (not mentioned in this text but can be reviewed in [103]) ensure that the photoreceptor cells are able to react to fast temporal changes in the visual signal as well. With increasing light stimulation or photoisomerization, the photocurrent increases as well until all CNG-channels are closed. At this point, the saturation state of the phototransduction is reached, leading to a maximum photocurrent of $\sim 13\,pA$ in a single human rod [116].

Due to the large amplification of the phototransduction, rods are able to detect single photons. Furthermore, they are still sensitive up to a photon quantity of 10^6. Cones that are somewhat less sensitive to photons, generate measurable potential changes only after several hundred absorbed photons. However, they can adapt quickly to new light conditions and thus scale their amplitude over an even wider range of intensities.

2.6.1.3 Horizontal Cells

As already mentioned in section 2.6.1, the horizontal cells are directly connected to the photoreceptor cells and situated in the INL. They possess relatively large receptive fields, i.e., they form synaptic connections to the underlying photoreceptors covering a huge area. In most mammalian, there are two different kinds of horizontal cells; one that connects selectively to cones and one that connects to both rods and cones. With their dendritic arbor, the horizontal cells detect the average intensity over their receptive field. Their axons are also connected to the photoreceptor cells. According to the average intensity detected by the dendrites, they generate an inhibitory signal, that is fed back to the photoreceptor cells via their axons. Thereby the horizontal cells enable that the retina can see at different intensity conditions with sufficient contrast.

2.6.1.4 Bipolar Cells

The photoreceptor cells react to light stimuli with a generator potential, i.e., the curve displaying the resulting potential change contains, in addition to the spectral information of the single cone types, information about the intensity and the temporal change, etc., of the visual signal. These different visual information need to be transferred to the brain via a series of action potentials, which always have the same amplitude and duration. The ganglion cells can only vary the frequency of those action potentials. This means that a generator potential of the photoreceptor cells needs to be translated into several binary signals. To realize this task, different bipolar cell types split the visual information into different channels, each of these channels is only sensitive to one specific feature of the visual signal.

To guarantee that every feature is detected for any visual stimulation, it needs to be ensured that every photoreceptor cell is connected to at least one of each bipolar cell types. The distribution of the bipolar cells across the retina is in that manner, that the whole area of the retina is covered completely by a specific bipolar cell type but there is no overlap in the receptive field of two bipolar cells of the same type [117]. This phenomenon is called tiling and applies to bipolar cells as well as to most ganglion, amacrine, and horizontal cells. However, the size of the dendritic arbors is not constant for one cell type. In the periphery, the bipolar cells connect to 5-10 cones and 20-80 rods [118]. The dendritic arbor becomes smaller near the fovea until there is a 1:1 connection at least for some bipolar cell types in the fovea itself [119]. Those bipolar cells that only connect to a single photoreceptor cell are called midget cells. They only exist in mammalians, that possess a fovea.

According to today's knowledge, there are around 13 different types of bipolar cells [119]. Until now, not all of those 13 different types have been correlated with a specific function. However, it is the common opinion that each of those types detects a specific feature in the visual signal. Only one of them detects the visual information of rods

[120]. The other $\sim$ 12 types are responsible for the transmission of the cone signal. Most of those types are non-selective to a certain cone type, which means they connect to all kinds of cones in their receptive field. So far we know of two bipolar cell types that are selective, one that connects selectively to blue cones [121] and one that connects selectively to green ones [122, 123].

Until now, there is no evidence for a red cone selective bipolar cell. Some theories argue that the red cones could only evolutionary survive in combination with the 1:1 connection of the fovea.

According to this wiring scheme, each cone in the fovea is connected to up to $\sim$ 11 different bipolar cells. How is it possible that those 11 different bipolar cells transmit different features of the visual signal if they all got the same input signal.

The answer lays in their receptors and the subsequent signaling cascade, which generate different responses to the input signal. In general, the bipolar cells can be subdivided into ON- and OFF-cells. OFF-bipolar cells exhibit receptors that lead to an opening of cation-channels during light and cause thereby depolarization of the membrane potential. The ON-bipolar cells on the other hand exhibit receptors, which cause a hyperpolarization of the membrane potential during light. Those two classes can be further subdivided whereas there are slightly more ON-bipolar cell types than OFF-bipolar cell types [117, 124]. For both classes, there are subtypes that react slowly on the input signal via metabotropic receptors and are therefore only able to detect low temporal frequencies (sustain-cells) and there are types that react quickly via ionotropic receptors and therefore can detect fast temporal changes in the visual signal (transient-cells). Additionally, there are bipolar cells that are sensitive to spectral contrast. Although the three cone types detect spectral information with different sensitivity, there are a variety combinations of intensities and spectral ranges, that could evoke the same reaction in a photoreceptor cell. Only a comparison of at least two different cone types can classify the signal into the visual spectrum.

All these different bipolar types evoke different membrane potentials for the same input signal, that is transmitted via their axon in the IPL to the ganglion cells.

A very helpful aspect for the investigation of those different bipolar cell types is, that different types not only distinguish in their gene expression and their function but in most cases also in their morphology. The most striking distinction thereby is their axon arbor, which not only differs in the degree of its branching but the different types also built their synaptic connection in characteristic depth of the IPL layer [118, 119, 125]. Thereby the IPL can be subdivided into 5 strata, i.e., layers, of equal thickness [126] (Fig. 2.12).

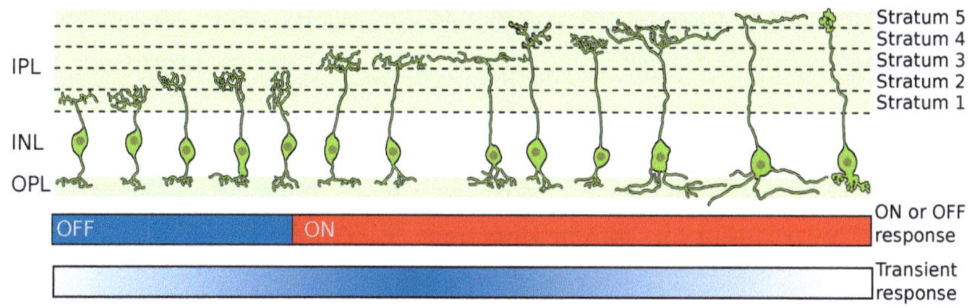

Figure 2.12: Morphology of the different bipolar cell types in mice, which are arranged according to their IPL stratification level. The figure was adapted from [119].

The OFF-bipolar cells, for instance, only stratify in the two outer strata one and two, while the ON-bipolar cells only stratify in the inner strata three, four, and five. Sustained bipolar cells stratify on the edges of the IPL, while transient bipolar cells tend to stratify in the middle of the IPL. Similar behavior is observed for cells that detect chromatic and achromatic information. [119]

2.6.1.5 Amacrine Cells

Just like the horizontal cells and the bipolar cells, the amacrine cells are located in the INL. The amacrine cells form both, their dendritic as well as their axial branches in the IPL. They detect the output signal of the bipolar cells and provide a corresponding feedback signal for a fine-tuning of the visual signal to be transmitted [127].

The interconnections and communication of the amacrine cells with the ganglion and bipolar cells can be very complex, but will not be discussed in the following.

2.6.1.6 Ganglion Cells

The ganglion cells are the last neuronal layer in the retina. As with the other cell classes, the ganglion cells can be subdivided into different types that are sensitive to specific characteristics of the visual signal. Currently, it is assumed that there are about 30 different types of ganglion cells [128].

Since the individual ganglion cells further refine the visual characteristics, that have already been split into different feature channels by the bipolar cells, certain ganglion cell types only form synaptic contact to certain bipolar cell types, which is why different ganglion cell branch their dendrites only in very specific strata in the IPL as well. The branching behavior in the IPL is therefore also an important aspect for the classification of different ganglion cell types.

Although there are still some gaps in the classification of ganglion cells, the previous classification has already provided some insights into information processing within this cell layer.

For example, several ganglion cell types are sensitive to a specific type of motion. At this point, there are several directional selective ganglion cell types identified [129–131] which are sensitive to motion, to specific directions [132, 133].

Additionally, it is known of some ganglion cell types, which are able to detect light by themselves [134–137]. It is assumed that some of these are involved in the circadian cycle. Furthermore, there are orientation sensitive cells [138–140], cells that react to chromatic contrast [141–147] and the already mentioned midget cells that are responsible for high acuity vision and make up the largest percentage of primate ganglion cell population.

Furthermore, the so-called local edge detector ganglion cells (LED-GC) have also been identified [140, 148–151]. In the following, the information transduction of the LED-GC is described in more detail. The name is a bit unfavorable because it describes a functional characteristic, which is common to a large number of the different ganglion and bipolar cell types. The LED-GC make up a large part of the ganglion cell population in rabbits and mice (13% and 15%, respectively) [152,153]. In primates, the proportion is lower, because of the high number of midget cells, but they are still one of the most common ganglion cell types. In contrast to other ganglion cells, they cover only very small fields with their dendritic branching (in mice about 100 µm) [153, 154], which stratifies in the middle strata of the IPL.

The functional characteristic, which gives the LED-GC their name, is that the information processing is not homogeneous in their receptive field. The receptive field describes the area of the photoreceptors that can influence the signal in a ganglion cell through the different interneurons (horizontal, bipolar, and amacrine cells). The receptive field of the LED-GC is of round shape and can be divided into two areas; a central and a surround area (Fig. 2.13). These areas act antagonistically to each other. This basic division of the receptive field have many cell types in the retina in common and will be explained in more detail here for the LED-GC. The LED-GC are sensitive to light falling in the center of their receptive field and are inhibited by light detected in the surround area. Therefore, the strongest output signal is generated in these cells when only the central area is stimulated, whereas no output signal is generated when the entire receptive field is stimulated [153]. Due to this property, they generate higher signals at edges of stimulated areas while they are not activated inside these areas.

Furthermore, LED-GC are sensitive to varying (phasic) signals, i.e., they generate a stronger output on rapidly changing signals [140, 155].

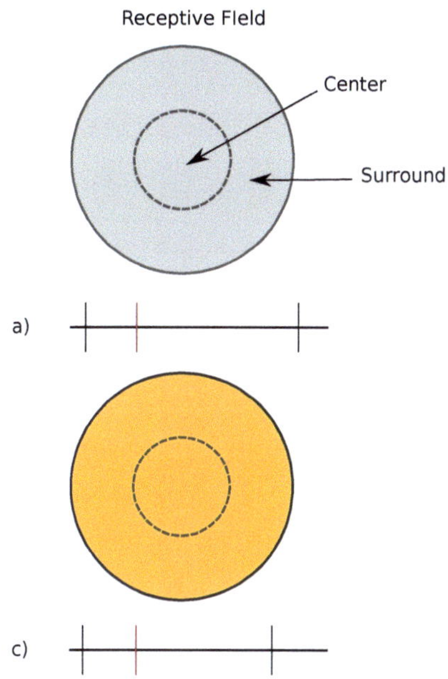

Figure 2.13: Schematics for information transduction within the receptive field of LED-GC. a) Receptive field of a LED-GC with its center and surround region. With no stimulation only a baseline of action potentials are generated. b) If only the center of the receptive field is stimulated the a high frequency of action potentials are generated. c) If the whole receptive field is stimulated the generated action potentials do not exceed the base line. d) If an edge is within the receptive field the LED-GC generate on moderate output signal.

This type of information processing ensures that an increased frequency of action potentials is generated especially when the signal changes (both temporally and laterally), while redundant information about constant signals does not generate any output signal, which leads to an efficient information compression of the visual signal.

3 Full-Field Swept-Source OCT

All measurements shown in this thesis are done with a full-field swept-source OCT (FF-SS-OCT) system. Due to the parallel illumination and detection of the field of view combined with a fast acquisition rate of the depth information, the system is able to detect whole retinal volumes within about 8 ms. Thereby it is possible to get rid of scanning artifacts and to reduce the influence of retinal motion. To this end, phase-locked data are generated, which are suitable for further post-processing. Beside, a numerical correction of dispersion, axial motion, and aberration, this enables detection of morphological changes in the sample with nanometer precision. The following chapter describes the setup as well as the post-processing necessary for detecting changes in the different neuronal layers in the retina after optical stimulation.

3.1 Setup

The general setup of an OCT system is an interferometer, where light from a light source is split into a sample and a reference arm. This can be done with a Michelson interferometer where the light is split into two beams by a semitransparent mirror and then again superimposed with the same mirror (Fig. 3.2). An alternative setup variant is a Mach-Zehnder interferometer, where the light is split and superimposed with two different components (Fig. 3.1). During the scope of this thesis, first, a setup was used, that was based on a Mach-Zehnder interferometer and later changed to a setup based on a Michelson interferometer. However, the same components and basic settings were used, so that the choice of the interferometer does not influence the results of the measurements. For most of the measurements in this thesis, the setup based on the Michelson interferometer was used. For measurements made with the Mach-Zehnder based setup, this is explicitly mentioned in the results. In the following, the Michelson setup is described, followed by a section on the Mach-Zehnder setup.

3.1.1 Mach-Zehnder-Based Setup

The Mach-Zehnder based setup (Fig. 3.1) is very similar to the Michelson interferometer setup described above. The basic components like the broadsweeper, the booster module, the high-speed area camera, as well as most other optical components are the same as described above. In contrast to the Michelson based setup, the light from the

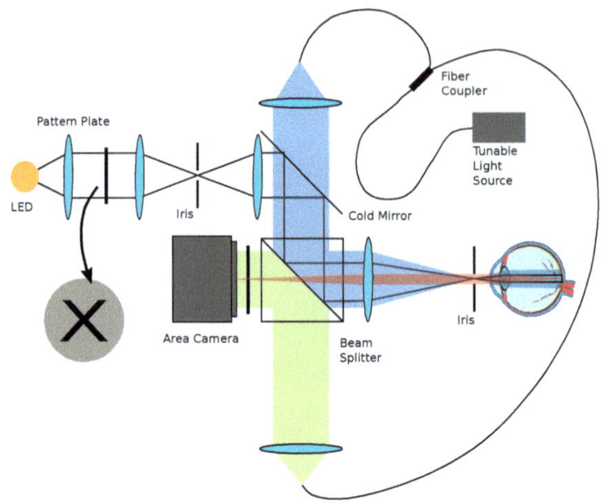

Figure 3.1: FF-SS-OCT setup: The light of the broadsweeper is split by a 90/10 fiber coupler into a reference illumination (green) and a sample illumination (blue). The superposition of the reference light and the light scattered back by the retina (red) is detected by an area camera. The stimulation (white) is coupled via a cold mirror into the sample arm to project an image of the x-shaped mask onto the retina.

booster module is split into two partial beams by a 90/10 fiber coupler. In doing so 90% of the light is introduced into the sample arm and 10% into the reference arm. The setup of the sample, as well as its image into the eye, corresponds exactly to the sample arm described above. The stimulation unit coupled into the sample arm is also identical. The light which is coupled into the reference arm is expanded to a beam profile of 20 mm diameter using a collimator (custom made). This collimated beam is then superimposed with the sample light scattered back from the sample by a beam splitter cube (Edmund Optics TECHSPEC Non-Polarizing Cube Beamsplitter, 50 mm NIR) and detected by the high-speed camera.

3.1.2 Michelson-Based Setup

For the Michelson FF-SS-OCT system (Fig. 3.2) a tunable light source (Superlum, Broadsweeper BS-840-01) with a central wavelength of $\lambda_0 = 841.5$ nm and a bandwidth of $\Delta\lambda = 51$ nm is used. Its output power of 3 mW is raised by an optical power booster (Superlum, optical power booster SOA-840-30) to 30 mW. The light source is connected via a fiber (M00299737, 780HP-Custom-Made) to a collimator (custom made), which expands the beam to a radius of 16 mm. A combination of cylindrical lenses (Thorlabs,

LJ1695L1-B, and LJ1567L1-B) is used to create an elliptic beam profile of radius of 32 mm × 16 mm. Since only the central, homogeneous parts of the beam profile should be used, the beam is restricted via a 20 mm×8 mm rectangular aperture. Afterward, the beam is divided into a reference and a sample arm by a 50/50 beam splitter (Edmund Optics TECHSPEC Non-Polarizing Cube Beamsplitter, 50 mm NIR). The light of the reference arm is focused via a suitable lens system onto a reference mirror, where it is reflected, whereas the sample arm illuminates an area of about 3.6 mm × 1.5 mm on the retina. On the retina, the total radiant flux of the illumination is reduced to about 5 mW, which is below the maximum permissible exposure. A detailed safety consideration according to the laser safety standard DIN EN 60825-1 for this setup can be found in [156]. The backscattered light of the retina is superimposed with the collimated reference light and imaged with a 5-times magnification onto a 2D camera (Photron, FASTCAM SA-Z). An iris in the focal plane of the eye limits the numerical aperture to ensure correct sampling of all lateral frequencies by the camera. A single volume is acquired by taking multiple images, while the light source is tuned linear by the wavenumbers from $2\pi/867\,\mathrm{nm}$ to $2\pi/816$ nm, this results in an axial resolution (full-width half maxima) of

$$\Delta z = \frac{1.21\pi}{\Delta k} = 8.4\,\mu\mathrm{m}$$

in air [82]. Here $\Delta k = 2\pi/816\,\mathrm{nm} - 2\pi/867\,\mathrm{nm}$ is the spectral bandwidth in wavenumber and 1.21 is a factor which is introduced by considering a rectangular shape of the spectra. The camera acquires 512 images of size 640 px × 368 px (corresponding to a field of view of 2.6 mm × 1.5 mm on the retina) with a frame rate of 60 000 Hz for each volume. Thereby, volume rates of 117 Hz can be reached. The data are stored with 8-bit precision on the internal storage of the camera during the measurements and are afterward transferred to the PC for further post-processing. With a raw data volume size of 640 px × 368 px × 512 px, 70 volumes can be acquired before the memory chip of the camera is exhausted. Thus, at full duty cycle, the retina can only be imaged for about 0.6 seconds. To increase the total measurement time a delay between successive volumes is introduced. The maximal delay, which can be used here is 125 ms. If the delay is increased further, motion between volumes becomes too severe, to co-register them effectively. For stimulation of the retina with bright white light, an LED (Roithner Lasertechnik GmbH, B5B-430-JB) is used (Fig. 3.3). Light of the LED is brought via a lens on an x-shaped aperture. The image of this pattern is then projected via a 4-f optics onto an image plane of the FF-SS-OCT setup, which is conjugated to the retina. An iris in the Fourier plane minimizes the influence of aberrations on the image of the pattern. A cold mirror (Edmund Optics TECHSPEC Cold Mirror, 45° AOI, 50.0 mm square) is used to couple the stimulation beam into the sample illumination arm of the setup. To ensure that no stimulation light disturbs the camera image, an infrared long-pass filter (Edmund Optics Optical Cast Plastic IR Longpass Filter, 4" × 5") is placed in front of the camera chip.

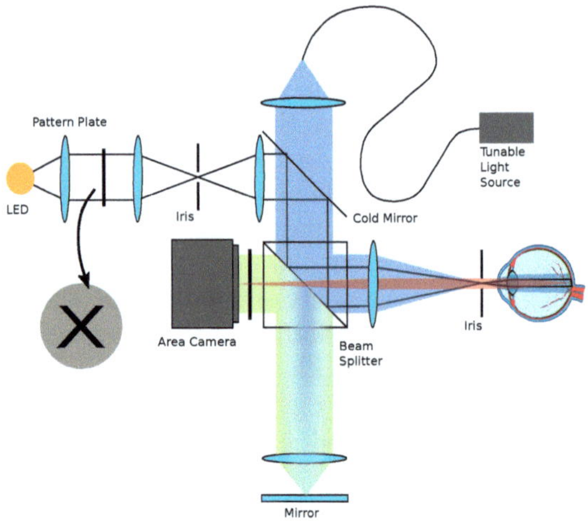

Figure 3.2: FF-SS-OCT setup with optical stimulation: The light of the broad sweeper is split by the beam splitter cube into a reference illumination (light blue/green) and a sample illumination (blue). The superposition of the reference light and the light scattered back by the retina (red) is detected by an area camera. The stimulation is coupled via a cold mirror into the sample arm to project an image of the x-shaped mask onto the retina.

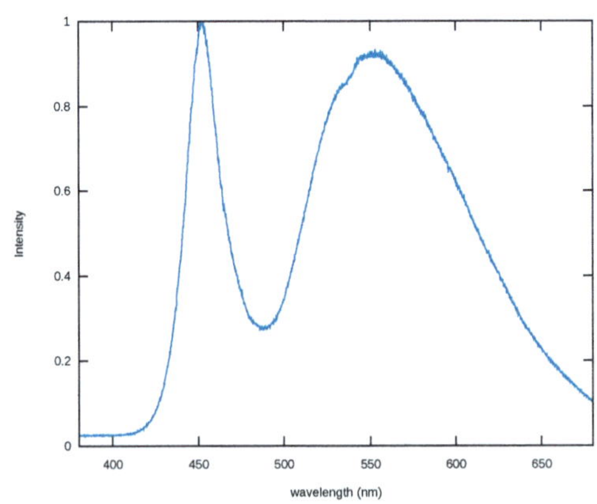

Figure 3.3: Intensity spectrum of the LED used for visual stimulation.

To synchronize the stimulus with the measurements, a microprocessor board (Arduino UNO) was used, which triggers the camera, the tunable light sources, and the LED.

3.2 Imaging Human Retina

The stimulus in each measurement is triggered with the 5th volume so that the first five volumes can be used as a baseline.

All measurements shown here were carried out on a single volunteer. Written informed consent was obtained from the subject. Compliance with the maximum permissible exposure of the retina and all relevant safety rules was confirmed by the responsible safety officer and the study was approved by the ethics board of the University of Lübeck.

The volunteer was female with no known previous diseases or ametropia. During the measurements, the volunteer aged between 26 and 29 years.

To improve the imaging, the pupil of the volunteer was artificially dilated with eye drops to a diameter of approximately 8 mm. Before the beginning of each measurement session, the test person was dark-adapted for at least 20 minutes, the measurements were also performed in a darkened room. After each measurement, a break of at least 5 minutes was taken until the next measurement to obtain a complete regeneration of the stimulated area in the retina.

3.3 Post-Processing

With FF-SS-OCT, the generation of a good and comparable image quality is not linked to high demands on the setup adjustment but the subsequent post-processing. The image quality of the generated raw data therefore seems rather poor compared to other OCT data. In order to obtain competitive images, other steps besides the reconstruction of the depth profile are necessary. Therefore, artifacts caused by lens reflexes and blurring of the image structures caused by phase errors have to be removed. Besides, evaluating the FF-SS-OCT volumes for functional analysis requires an additional high level of post-processing.

Global motion, that occurs in between volumes, needs to be removed to ensure that the exact same position of the retina is compared over time. Additionally, the structures of interest need to be segmented with high reproducibility. Since the phases of OCT data are sensitive to nanometer-motion which is below the OCT resolution, both registration, as well as segmentation, have to be done with sub-pixel precision. But even if this is done perfectly statistically independent speckle will emerge over time, due to metabolic processes in the retina. Therefore, a special phase evaluation is used, which provides useful phase information beyond the decorrelation time of a speckle pattern.

Thus, the evaluation has to face the challenges of poor image quality and high noise levels of FF-SS-OCT data. Since 4D data are acquired, huge amounts of data have to be handled. Therefore, algorithms that are robust against low image quality and high noise are needed. Additionally, those algorithms need to provide an efficient computation time. In the following section, the reconstruction, the correction of axial and lateral phase errors in the data, the registration, segmentation, and phase-evaluation algorithm which were used to face those challenges, are described.

3.3.1 Reconstruction

The reconstruction of the acquired data is done by a Fourier-transformation of the axial direction (Fig. 3.4), which transforms the interferograms into depth information. Those reconstructed images contain a lot of artifacts, which are mainly caused by reflections in the reference and sample arm. Those artifacts are static for different volumes, while the retina is in constant motion during the measurement. The average of several volumes of one measurement will therefore only contain the fixed noise, while the signal of the retina cancels each other out over time. To remove those artifacts a coherently averaged volume is subtracted from each volume.

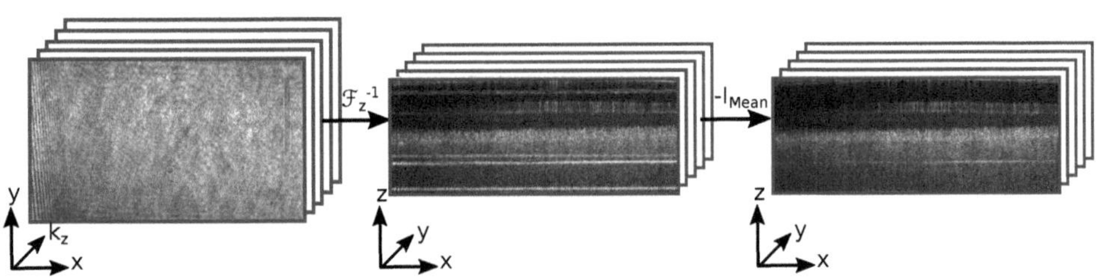

Figure 3.4: Reconstruction steps for FF-SS-OCT data. Raw camera images (left) are Fourier-transformed in the axial dimension to obtain the depth profile (middle) and the complex average of all volumes in one measurement is subtracted from each volume to remove lens reflexes.

3.3.2 Correction of Dispersion and Axial Motion

As already shown in section 2.4.4, dispersion mismatches in reference and sample arm as well as axial motion during the acquisition of a volume can cause an axial phase error function which is superimposed with the spectral information of the depth profile. This leads to an axial blurring of the volumes. In order to correct this, the phase error has to be determined first and then to be removed by multiplying the spectrum with the complex conjugated phase error. This is done via a deterministic approach, which retrieves the phase error function by using the correlation of sub-spectra. The following algorithm is described in details in the publication of Hillmann et al. [67].

To determine the phase error function the interference spectra $I_{\mathrm{disp}}(k)$ from Equation (2.17) are multiplied with a window function $w(k)$. The window function is chosen narrow enough, so that the present phase error can be approximated by a polynomial of first degree, which corresponds to an axial displacement of the structure (Fig. 3.5). While the center of the window function is shifted to different wavenumbers k_0' of the spectra the local gradient of the phase error function changes and correspondingly the image of the structure is axially displaced by $\Delta z(k_0') \approx \partial\Phi(k)|_{k=k_0'}$. The axial displacement is then determined by cross-correlation of the different reconstructed images.

The phase error function $\Phi(k)$ is then calculated by an integration of the local slopes

$$\Phi(k) = \int \mathrm{d}k\, \Delta z(k) + C.$$

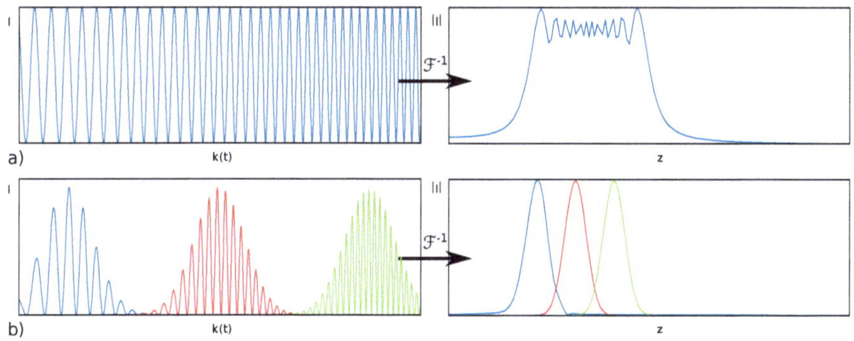

Figure 3.5: a) Interference spectrum of a single reflecting layer superimposed with a quadratic phase error function corresponding to an axial blurred profile. b) Reconstructions of the spectrum multiplied with three narrow window-functions centered at different wave numbers results in a better axial resolution of the signals in different depth. The axial displacement of the signals correspond to the first derivative of the phase error of the chirped signal.

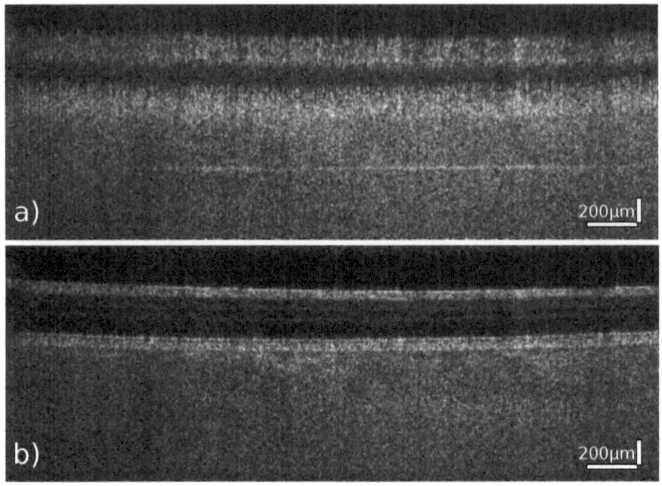

Figure 3.6: FF-SS-OCT B-scan of a healthy volunteer 2 mm in the periphery. a) After the reconstruction and removal of reflexes but without further post-processing. b) The same B-scan with correction of axial motion and dispersion.

The integration constant C is undetermined but does not influence the image intensity.

This way, the phase function can be retrieved by correlating a specific sub-spectra only to its adjacent sub-spectra. Yet, two different sub-spectra create statistical independent speckle in the images. Hence, the shift between those images can only be determined with an uncertainty, which depends on the speckle size. To reduce the influence of the speckle noise, each sub-spectral image is correlated with all other sub-spectral images, which results in an overdetermined phase error, which is then finally determined with a least-square fit. The result of the motion and dispersion mismatch correction is shown for a typical B-scan in Fig. 3.6.

3.3.3 Correction of Aberrations

The problem of aberrations is very similar to the previous problem of unmatched dispersion and axial motion. However, not an axial phase error function but a two-dimensional lateral phase error function $\Phi(k_x, k_y)$ must be compensated in the aperture plane.

To identify this phase error, an optimization and a deterministic approach for determine the phase error from the measurements are used here. In the first step the optimization is used to correct for defocus [68] and after that, a correlation of subapertures is used to determine higher-order aberrations [69].

Optimization of image quality The main idea of optimization is that several solutions for the posed problem are generated in a chosen parameter space. Those solutions are then evaluated by a previously chosen criterion. The parameter space of the solution is then varied in a manner, that it converges to an optimal solution regarding the chosen criterion. This means that a parameter space is needed that can describe the complex conjugated phase error function with as few parameters as possible and a criterion that judges how well the existing phase error is compensated by a given phase function.

In order to parameterize aberrations the Zernike polynomials, Z_n^m are often used. The Zernike polynomials are a series of orthogonal polynomials defined on the unit disc. The single Zernike polynomials are thereby defined as follows

$$\begin{aligned} Z_n^m &= R_n^m(r)\cos(m\theta) \\ Z_n^{-m} &= R_n^m(r)\sin(m\theta) \end{aligned} \tag{3.1}$$

Here m and n are non-negative integers, r describes the radius, and θ the azimuth angle of the polynomial. The radial polynomial R_n^m are defined as:

$$R_n^m(r) = \begin{cases} \sum_{k=0}^{\frac{n-m}{2}} \frac{(-1)^k (n-k)!}{k!\left(\frac{n+m}{2}-k\right)!\left(\frac{n-m}{2}-k\right)!} r^{n-2k} & \text{for}\, n-m \,\text{is even} \\ 0 & \text{otherwise.} \end{cases} \tag{3.2}$$

In Figure 3.7 the corresponding Zernike polynomials up to the 5th radial degree are shown. Since in the first step only the defocus of the volume is corrected by an optimization, only the Zernike Polynomial Z_2^0 is used, whereby two parameters are optimized; the offset and the slope of the defocus, as a function of depth z. By doing so the defocus is not only corrected for a single plane in the volume but all planes in the volume. The second part of optimization is to evaluate how well the existing phase error can be compensated a defocus. Here the Shannon entropy is used for this purpose, which is defined by

$$S = -\sum_{(x,y)} I \log(I). \tag{3.3}$$

Here, I is the intensity of the complex OCT image. This particular formulation of Shannon entropy weights areas of low intensity very high and thus prefers the emergence of dark areas in the image. This gives phase functions, which cause high contrast small values in the metric function. On the other hand, the high weighting of high

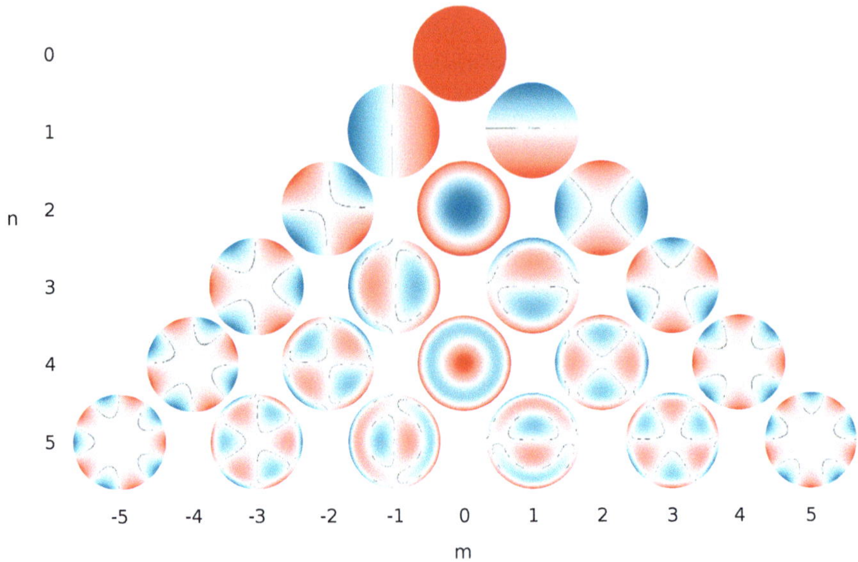

Figure 3.7: Zernike polynomial up to the $n = 5$ order.

intensity would not work because they cannot be limited upwards and thus cannot prevent that the erroneous accumulation of the intensity of different scatterers will create better metric value than the real sharp image.

Since optimization can have the risk of running into a local minimum instead of a global minimum of the metric function , the aperture is first reduced numerically, i.e., a small area in the middle of the aperture plane is cut out (Fig. 3.8 a-c). With a smaller aperture, the depth of focus is higher, and aberrations have less influence on the image sharpness and therefore the starting point is closer to the desired result.

Correlation of Sub-Apertures When an optimal compensation for the defocus has been found for the small aperture, higher-order aberrations are corrected for this small aperture. To prevent the risk of running in a local minimum instead of finding the global minimum higher-order aberrations are not corrected with optimization, but with a correlation of sub-spectra. This approach is very similar to the previously describe dispersion correction. Small sub-apertures are cut off from the aperture plane or the lateral frequency spectrum of the volume (Fig. 3.9). If the sub-apertures are small enough the phase error within the sub-aperture can be approximated by a linear phase function. An inverse Fourier-transform of each sub-aperture yields an image, which is laterally shifted according to the local phase gradient present in the sub-aperture. The

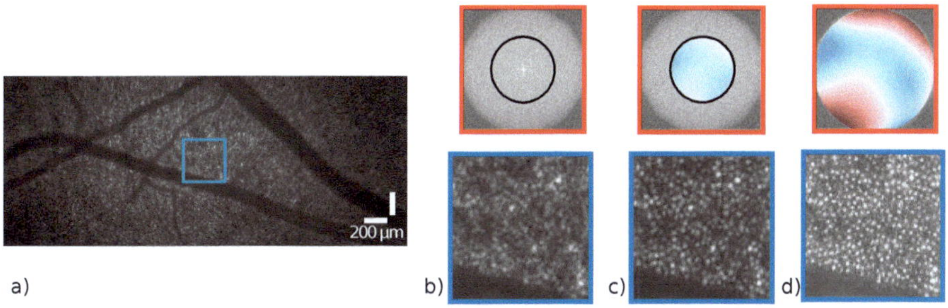

Figure 3.8: Effects of aperture size on the aberration correction. The blue marked rectangular region is shown in close up in b-d). b) The region without aberration correction. c) The region after the defocus and other higher aberration are corrected for a small central aperture. d) The results when the defocus and the aberration for the whole aperture are corrected.

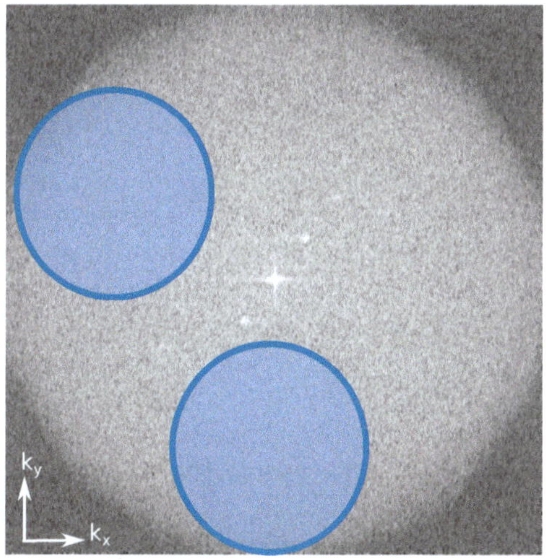

Figure 3.9: Wavefield reconstructed in the aperture plane with two random artificial sub-apertures.

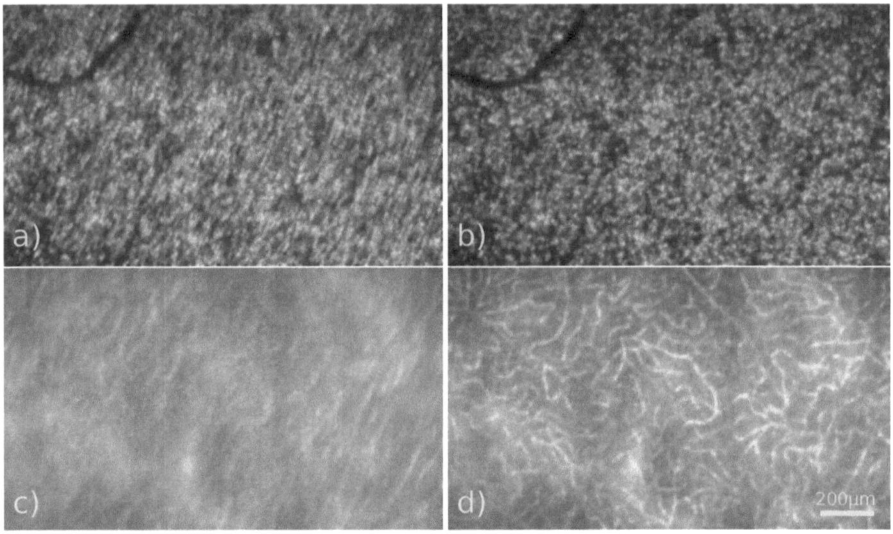

Figure 3.10: En-face images of a FF-SS-OCT measurement in 3 mm temporal, averaged over 70 volumes after co-registration. a) The photoreceptor layer before an aberration correction. b) The same layer after the here described numerical aberration correction. c) The IPL of the same measurement before the aberration correction, while d) displays the IPL after the aberration correction.

cross-correlation of these different sub-aperture images gives their local displacement $\Delta \boldsymbol{x}$, which is proportional to the local phase gradient $\nabla_k \Phi = \left(\frac{\partial}{\partial k_x}\Phi, \frac{\partial}{\partial k_y}\Phi\right)$ between these images. A correlation of several sub-apertures across the aperture plane can thereby approximate the present phase error function. Again, as for the dispersion correction, the speckle noise of the images leads to uncertainty, limiting the accuracy, by which the lateral shift can be determined. To reduce the influence of the speckle pattern, an overdetermined system is created, by correlating each sub-aperture image to several other randomly chosen sub-aperture images. In order to obtain the lateral phase error function (amplitude transfer function) a linear combination of several Zernike polynomials are fit to the thereby retrieved phases via a least-square fit [69].

When the higher-order aberrations are corrected, then the radius of the aperture is increased and again the defocus is first corrected via optimization and the higher-order aberration via the correlation of the sub-apertures. This is done iteratively until the aberrations for the complete aperture are corrected (Fig 3.9 d).

The results of the aberration correction are shown in Fig 3.10. One advantage that a numerical aberration correction offer is that those corrections can be done for different depth of one volume by simply applying different correcting phase functions in the aperture plane, i. e., in contrast to adaptive optics, that are only able to focus on a single depth at a time, whole focused volumes can be obtained with a numerical aberration correction.

3.3.4 Registration

During the measurement, there is a continuous movement of the eye. These movements are mostly involuntary and therefore cannot be suppressed by the test person. Since the FF-SS-OCT setup described here is not able to track these movements directly during the measurement and to correct the illumination of the retina accordingly, each volume of a time series is slightly shifted to the previous volume. The movements that occur can be divided into three different categories, which differ in frequency, speed, and amplitude and hence have different effects on the resulting data. The following characterization of these eye movements is based on [157, 158].

- **Tremor** A tremor in the eye is caused by an inaccuracy in muscle control. The movements caused by tremors are very small, aperiodic, and high-frequent. They have an amplitude of 1.5 µm – 5 µm at a maximal frequency of 150 Hz. With their amplitude, the movements are close to the resolution limit of the FF-SS-OCT setup used. Thus, they may still be sufficient to shift the speckle patterns significantly against each other.

- **Drift** Drifting of the eye takes place to avoid adaption of the retina to the visual stimulus. For this purpose, the image on the retina is permanently shifted by the drift, so that the photoreceptors and thus the rest of the retina is exposed to permanently changing visual impressions. The drift is very slow, with a speed of about 40 µm/s. They are fairly monotonous and can cause offsets of several dozens of micrometers. The drift is corrected by occasionally occurring microsaccades.

- **Microsaccades** Microsaccades are used to compensate for the drift of the retina and to realign the focus position. They occur at a rate of 1-4 times per second and can cause retinal displacements of up to 300 µm. They are very fast, reaching speeds of 90 mm/s. Usually, these microsaccades last 20 – 200 ms. Therefore, if a microsaccade occurs during the acquisition of a volume, the complete image information of this volume is destroyed, and it is useless for further analysis.

For the functional evaluation, the phase difference between volumes recorded at different times must be formed at a later point in post-processing. Eye motion leads to a phase difference calculated between different positions of the image field, i.e., different positions of the speckle pattern. This causes random phase differences, which

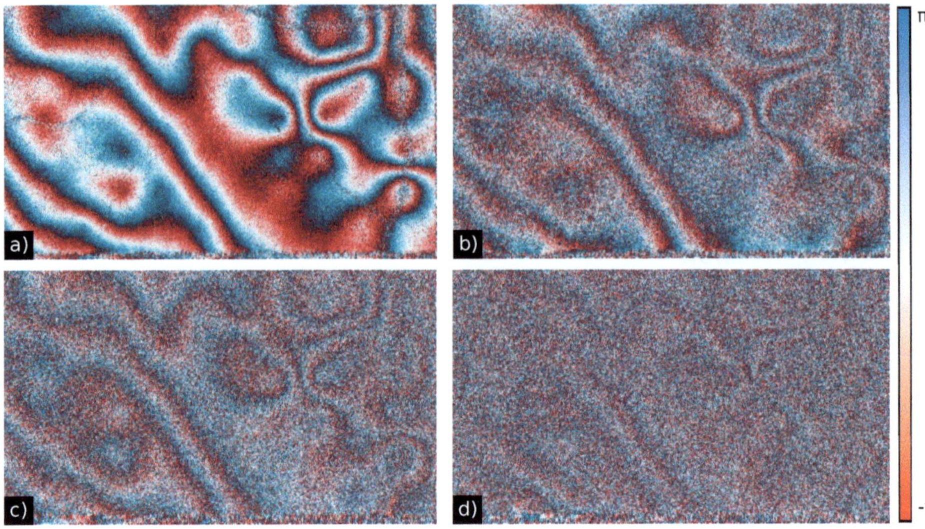

Figure 3.11: Temporal phase differences of laterally shifted volumes. a) Temporal phase difference of two volumes which are exactly registered. b) Temporal phase difference of the same two volume one of them is lateral shifted about 1 pixel. c) Temporal phase difference of the same two volume one of them is lateral shifted about 2 pixels. d) Temporal phase difference of the same two volume one of them is lateral shifted about 3 pixels.

no longer contain information about the sample. The task of the registration is, thus, to identify the movement that has taken place between the different volumes and to correct it accordingly. Once this is done, the phase difference corresponds to a lateral phase function, which is caused by movements of the pupil during the measurement, and or residual axial motion (Fig. 3.11 a).

How sensitive the phase is to even minimal shifts of the different volumes can be seen from this phase function. If the registration was exact and the speckle patterns of two different volumes were perfectly superimposed, then the phase difference of the two volumes corresponds to a smooth function (Fig. 3.11 a). But if the volumes are only minimally shifted then there are individual points in the speckle pattern for which the phases of the two volumes are no longer correlated and therefore their phase difference is random, while this is not yet the case for other points. This leads to a noisy phase difference, until eventually at a too large offset nearly only noise is present in the phase differences (Fig. 3.11 d). Since the average size of speckle for the imaging setup used here is about 1.2 pixels wide, even a shift of only one pixel leads to a significant noise level in the phase differences (Fig. 3.11 b). A requirement of the registration is, therefore, to correct the movements within a measurement with a sub-pixel precision to ensure a stable phase evaluation.

Another challenge is the handling of volumes destroyed by microsaccades. In order to correct the movements, one reference position needs to be selected and the offset of the different volumes in relation to this position has to be determined. By means of appropriate transformations, all volumes are then aligned congruent to the reference position.

In general it would be sufficient if the translation between two adjacent volumes is calculated and corrected. Thus, the reference volume is always shifted to the last volume which has been aligned. Nevertheless, if a microsaccade occurs during the measurement of a volume no meaningful correlation between the corrupted and the following volume can be obtained. Consequently, the translation after a microsaccade is random, which leads to a discontinuity after every saccadic motion. If, on the other hand, the displacement for each volume is only calculated to one fixed and predetermined reference volume one can avoid those discontinuities but the registration fails completely, if the reference volume is corrupted by a saccadic motion.

Thus, an important task of the registration is to identify such volumes which have no meaningful correlation to the other volumes and ensure that their influence on the registration is as small as possible.

3.3.4.1 Determination of the Movements

In order to meet the above mentioned requirements, an overdetermined system of correlation-based information is generated for the determination of eye movements. The basic idea of this determination will be explained in the following. For registration of large data sets cross correlations are often used, because they are particularly computationally efficient compared to other registration methods. With cross-correlation the correlation of volumes is calculated for different displacements of the volumes to each other. If the volumes are only shifted against each other, then the cross-correlation receives its greatest value at exactly the position by which the two volumes are shifted against each other. This can also be realized in frequency domain according to the Fourier-shift theorem and the convolution theorem for Fourier-transforms. If it is assumed, that $I_i(\boldsymbol{x})$ and $I_j(\boldsymbol{x})$ show the same image only that $I_j(\boldsymbol{x})$ is shifted by $\Delta\boldsymbol{x}_{i,j}$, i.e. $I_j(\boldsymbol{x}) \approx I_i(\boldsymbol{x} - \Delta\boldsymbol{x}_{i,j})$ applies, then, according to the Fourier-shift theorem, the spectrum of $I_j(\boldsymbol{x})$ is only multiplied by an additional phase factor, as followed:

$$\mathcal{F}\{I_j(\boldsymbol{x})\} \approx \mathcal{F}\{I_i(\boldsymbol{x} - \Delta\boldsymbol{x}_{i,j})\} = \tilde{I}_i(\boldsymbol{k})\exp(-\mathrm{i}\boldsymbol{k} \cdot \Delta\boldsymbol{x}_{i,j}). \tag{3.4}$$

According to the convolution theorem, the spectrum of the cross-correlation $\tilde{\Gamma}_{cc}(\Delta\boldsymbol{x})$ can then be calculated by

$$\tilde{\Gamma}_{\rm cc}(\Delta \boldsymbol{x}) = \tilde{I}_i(\boldsymbol{k}) \cdot \tilde{I}_j(\boldsymbol{k})^* \approx |\tilde{I}_i(\boldsymbol{k})|^2 \exp(+\mathrm{i}\boldsymbol{k} \cdot \Delta \boldsymbol{x}_{i,j}), \tag{3.5}$$

where * denotes the complex conjugate. By doing so the phase of the original spectrum is removed and only the phase factor remains that describes the offset of the two volumes to each other. A transformation back into the spatial domain then leads to a delta peak $\delta(\boldsymbol{x} - \Delta \boldsymbol{x}_{i,j})$ folded with a PSF, which is derived from the intensity of the spectrum, $|\tilde{I}_i(\boldsymbol{k})|^2$. The convolution results in a smearing of the information and to a deterioration of the SNR, and therefore this leads to an uncertainty with which the desired offset can be determined. If the two volumes are additionally overlaid with uncorrelated noise, the inaccuracy is even greater. However, the intensity $|\tilde{I}_i(\boldsymbol{k})|^2$ of the correlation spectrum does not carry any information about the offset itself but about the image structure and how well the two volumes can be correlated for different offsets, which means that this part of the cross-correlation can be removed by normalization in the frequency domain, according to

$$\tilde{\Gamma}_{\rm pcc}(\boldsymbol{k}) = \frac{\tilde{I}_i(\boldsymbol{k}) \cdot \tilde{I}_j(\boldsymbol{k})^*}{|\tilde{I}_i(\boldsymbol{k}) \cdot \tilde{I}_j(\boldsymbol{k})^*|} \approx \exp(-\mathrm{i}\boldsymbol{k} \cdot \Delta \boldsymbol{x}_{i,j}). \tag{3.6}$$

Such a correlation $\tilde{\Gamma}_{\rm pcc}$ is also called phase correlation, as only the phase of the cross-correlation is remaining. As the bandwidth of the spectrum is not infinite, the convolution with a PSF cannot be avoided completely. Yet, since the shape of the PSF can be estimated here, a fit function in form of the PSF to the maximum and the neighboring points can be used to determine the position $\Delta \boldsymbol{x}_{i,j}$ of the maximum with a sub-pixel accuracy.

In general, this offset must be calculated for each volume only to a previously determined reference position or a reference volume, but even with such a phase correlation the offset can only be determined with a certain error, because of superimposed speckle patterns and uncorrelated noise on the different volumes. Furthermore, the presence of other transformation such as rotations and deformations of the volumes lead to an additional uncertainty.

To maximize the accuracy an overdetermined system is generated from correlating all volumes of a measurement. I.e., each volume is correlated with $N-1$ other volumes, where N is the number of volumes in one measurements. While the actual distance to the reference position is the same for each of the $N-1$ offsets, the error for each offset is random. Thus by averaging over several offsets the error can be minimized. By correlation of each volume to every other volume $N(N-1)/2$ displacements $\Delta \boldsymbol{x}_{i,j}$ are obtained. While only one displacement to a reference volume $\Delta \boldsymbol{x}'_i$, $i = 0, \ldots N-1$ for each volume i is needed to correct the movements. This establishes the following system of equations:

$$\Delta \boldsymbol{x} = M \Delta \boldsymbol{x}' = \begin{pmatrix} \Delta \boldsymbol{x}_{0,1} \\ \Delta \boldsymbol{x}_{0,2} \\ \vdots \\ \Delta \boldsymbol{x}_{1,2} \\ \vdots \end{pmatrix} = \begin{pmatrix} 1 & -1 & 0 & \cdots & 0 \\ 1 & 0 & -1 & \cdots & 0 \\ & & \vdots & \ddots & \\ 0 & 1 & -1 & \cdots & 0 \\ & & \vdots & & \vdots \end{pmatrix} \begin{pmatrix} \Delta \boldsymbol{x}'_0 \\ \Delta \boldsymbol{x}'_1 \\ \vdots \\ \Delta \boldsymbol{x}'_{N-1} \end{pmatrix} \tag{3.7}$$

Here, the matrix M describes which volumes were used to calculate the corresponding offset $\Delta \boldsymbol{x}_{i,j}$. Since the offset $\Delta \boldsymbol{x}'$ is sought here, this is an inverse problem. The different offsets $\Delta \boldsymbol{x}_{i,j}$ between the different volumes can only be determined within a certain error, therefore there is no unique solution for this system. It is thus an ill posed inverse problem. In order to still obtain a solution for the displacement $\Delta \boldsymbol{x}'$ to the reference volume, the system is solved by a Tikhonov regularization. For this purpose the problem is reformulated as a least square problem, according to:

$$\|\Delta \boldsymbol{x} - M \Delta \boldsymbol{x}'\|^2 + \lambda \|\Delta \boldsymbol{x}'\|^2 \longrightarrow \min_{\Delta \boldsymbol{x}'} \tag{3.8}$$

Here λ is a regularization parameter.

Nevertheless, by determining the displacement $\Delta \boldsymbol{x}'_i$ for a volume i, via this system, all displacements $\Delta \boldsymbol{x}_{i,\cdot}$ of the measurement are included to the same extend. Yet, different correlation also produce results of varying accuracy. For example, for volumes that are rotated against each other in addition to a displacement, the error is larger than for volumes, that are only translated. Very large errors can occur if a volume has been destroyed by a saccade, because it no longer has a meaningful correlation to the other volumes. To minimize the error more efficiently a weight is needed, which describes the quality of the correlation.

As mentioned before, the magnitude of the correlation spectrum indicates how well two volumes can be correlated by a shift. Therefore, this value is suited to be used as a weighting for the individual correlations to reduce the influence of bad correlations on the final result. The formulation of the least square problem changes accordingly:

$$\|W(\Delta \boldsymbol{x} - M \Delta \boldsymbol{x}')\|^2 + \lambda \|\Delta \boldsymbol{x}'\|^2 \longrightarrow \min_{\Delta \boldsymbol{x}'} \tag{3.9}$$

Here W is a diagonal matrix containing the weights for the respective correlation. The solution of this least square problem can be computed to

$$\Delta \boldsymbol{x}' = (M^T W^2 M + \lambda E)^{-1} M^T W^2 \Delta \boldsymbol{x}, \tag{3.10}$$

where E stands for the identity matrix and ^{T} for the respective transposed matrix. By using $N-1$ different correlations to determine the offset of a single volume from the reference volume, and additional weighting them according to their accuracy the actual offset is determined with an increased accuracy.

3.3.4.2 Algorithmic Implementation

The algorithmic implementation of the registration is divided into two parts. The first part is a rigid registration and corrects only the rough displacements in x, y and z direction, while the second, non rigid part of the registration also corrects smaller movements, as well as rotations and deformation of the volumes against each other.

Rigid registration The coarse, rigid registration is divided into a lateral and an axial registration. This allows the use of projections of the individual volumes in the direction perpendicular to the registration direction, which improves the image quality and reduces the noise level.

Basically, there are two different lateral structures in each volume. First, the lateral structure of the retina and second, an intensity distribution superimposed on it, which is caused by the illumination of the FF-SS-OCT system. This intensity distribution is static and would therefore provide incorrect information for cross correlation.

Therefore, in order to create a projection $P_z(x, y)$ for the lateral registration, all en-face images of a volume are summed up to obtain V_{sum_z} in a first step according to:

$$V_{\text{sum}_z}(x, y) = \sum_{z=0}^{N_z} |V(x, y, z)| \tag{3.11}$$

Here $V(x, y, z)$ describes an individual volume of the measurement and N_z is the number of pixels in z-direction. The resulting en-face image V_{sum_z} contains only the information of the illumination pattern, while the information of the lateral structure of the retina is canceled out. In order to remove the illumination pattern, the volume is normalized with this sum.

$$V'(x, y, z) = \frac{V(x, y, z)}{V_{\text{sum}_z}(x, y)} \tag{3.12}$$

The axial projection $P_z(x, y)$ is then obtained according to:

$$P_z(x, y) = \left| \sum_{z=0}^{N_z} V'(x, y, z) \right|^2 \tag{3.13}$$

Afterwards, the projection $P_z(x, y)$ for each volume contain lateral structures of the retina such as vessels that can be used for the phase correlation.

With the different projections of the individual volumes, an overdetermined system of lateral phase correlation is then established as described in section 3.3.4.1, which is solved with the Tikhonov regularization. Therefore the first volume of the measurement is used as reference volume. The calculated offsets $\Delta\boldsymbol{x}_n = (\Delta x_n, \Delta y_n)$ for the different volumes V_n are then corrected via the Fourier-shift theorem, by multiplication of a respectively phase factor in the lateral Fourier domain, according to.

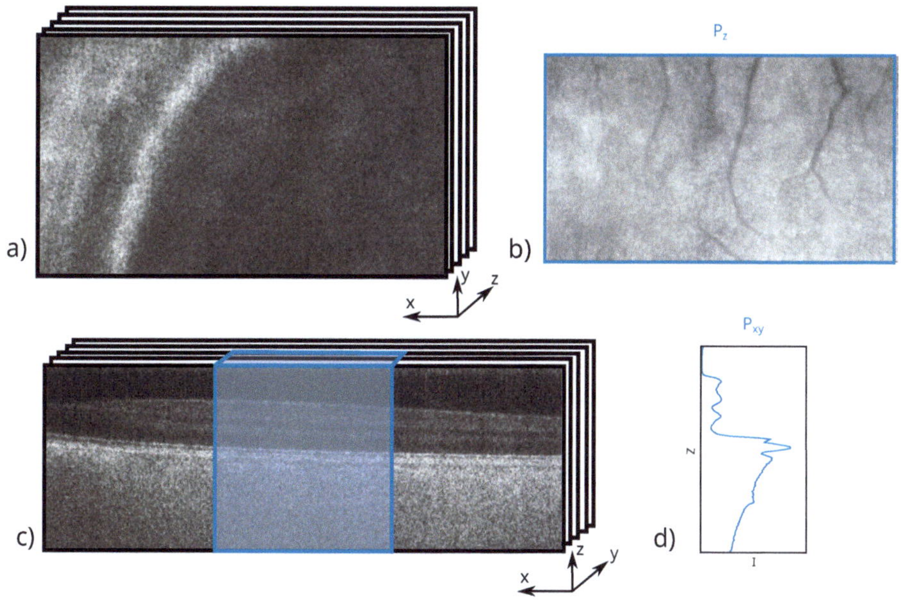

Figure 3.12: Generation of the axial and lateral projection for the registration. a) Different en-face images of a volume are averaged in z-direction to generate an axial projection of the important lateral structures of the volume to improve the lateral registration. b) Exemplary axial projection. c) B-scan of a volume. The blue marked region display the area in which an averaging in x- and y-dimension is done for the lateral projection. d) Exemplary lateral projection for a better axial registration.

$$V_n^{\mathrm{corr}} = \mathcal{F}_{x,y}^{-1}\left\{\mathcal{F}_{x,y}\{V_n\}\mathrm{e}^{-\mathrm{i}(k_x\Delta x_n + k_y\Delta y_n)}\right\} \tag{3.14}$$

Here V_n^{corr} represents the $n-$th volume after correction of the offset.

For the axial correlation of the different volumes, a projection $P_{x,y}(z)$ is created for each individual volume. Therefore a mean A-scan is generated by averaging the axial information of different positions (x, y). Since the retina in the OCT volume is slightly curved, averaging all lateral positions of a volume would result in a drastically smeared A-scan. For this reason, only a central area of 32×32 pixels is used to create the average A-scan, for which it can be assumed that the retina is more or less straight. Afterwards, again according to section 3.3.4.1 an overdetermined system of

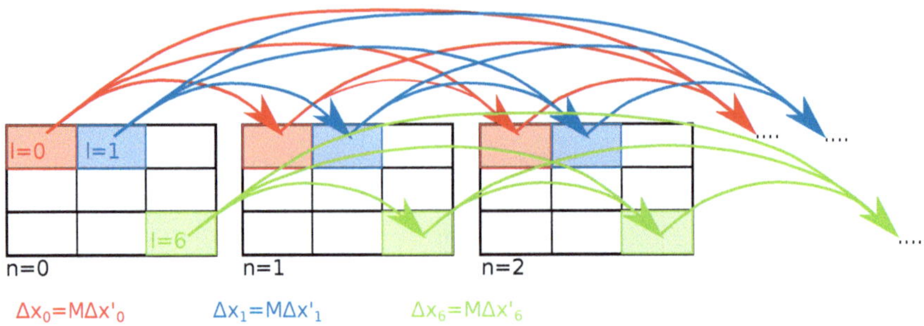

Figure 3.13: Illustration of establishing several overdetermined systems for different patches. The projection for each volume n is subdivided into several patches l (displayed in different colors). For each patch an own overdetermined system is established by phase correlations with the equivalent patch position in the other projections.

axial correlation is established and solved. Thereupon the volumes are shifted with the Fourier Shift theorem in axial dimension, according to:

$$V_n^{\mathrm{corr}} = \mathcal{F}_z^{-1}\left\{\mathcal{F}_z\{V_n\}\mathrm{e}^{-\mathrm{i}k_z\Delta z_n}\right\} \tag{3.15}$$

Non-Rigid Registration Also the non-rigid registration is divided into a lateral and an axial registration. At first, the lateral registration is performed. Therefore the lateral field of view, which is common to all volumes after the coarse registration, is selected. Only this field is used for the following spatially dependent, lateral registration. As already described above in Section 3.3.4.2, an axial projection $P_z(x, y)$ of this area is created for each volume. This projection is then divided into small tiles of the size 96×96 pixels in that manner, that there is at least an overlap of 24 pixels with the adjacent tiles.

Then, again an overdetermined system of phase correlation is established for each tile position by correlating the tiles of a given position from different volumes or projections (Fig. 3.13). This gives each tile l of each volume n an individual offset $\boldsymbol{\Delta}_{l,n} = (\Delta x_{l,n}, \Delta y_{l,n})$.

The different tiles within a projection can have very different offsets $\boldsymbol{\Delta}_l = (\Delta x_l, \Delta y_l)$, for example due to rotations or deformations.

In order to correct for such rotations or deformation, a two dimensional polynomial of 4-th degree is fitted to the different offsets within a projection to describe the present lateral deformation of the volume (Fig. 3.14 a). Due to the different offset within a

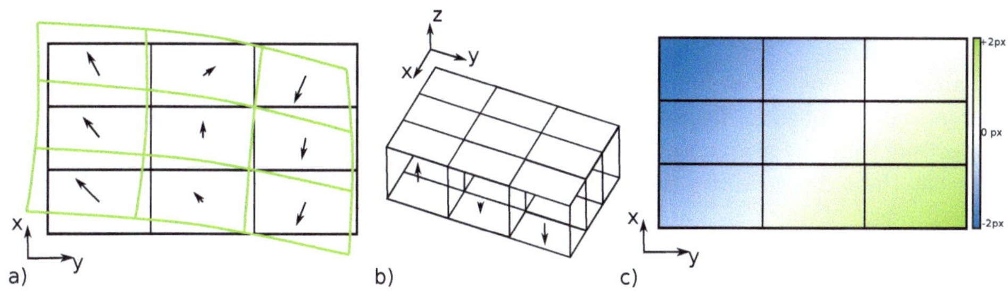

Figure 3.14: Illustration of the polynomial fit for the non-rigid registration. a) The different patches within one axial projection receive different lateral displacements. The resulting polynomial fit (green) corrects also for lateral rotations and deformation. b) During the axial correlation, the volume is again subdivided in several lateral patches, which each receive an axial displacement. Shown for the first row of the volume. c) The resulting polynomial fit corrects for axial rotation and deformation.

volume a grid with non equidistant nodes is thereby generated. Therefore, the single pixels of a volume can not simply be shifted according the fit function to correct the deformation in between two volumes. In order to correct such deformation correctly the image information of different pixels needs to be interpolated. This is done by using a two dimensional Gaussian kernel, which is convoluted with the image information.

After the volumes have been aligned with each other in accordance with the fine lateral registration, the last step is a fine axial registration. The non-rigid, axial registration is very similar to its lateral counterpart. Again, the axial registration takes place only in the lateral image field, which is common to all volumes of a measurement, like the lateral correlation before.

A lateral projection takes place to reduce the noise. Though, not one but a lot of lateral projection are created for a single volume. Therefore the field of interest is divided into (x, y) tiles of size 10×10 pixels, but this time, the tiles do not overlap. For each tile a lateral projection is calculated similar like the one described above. For each tile position of a volume, an overdetermined system of axial phase correlations with the corresponding A-scans of the other volumes is created and solved. This results in l different offsets Δz_l for the different tiles of one volume. Due to axial rotation and deformation, the different axial offsets can also be very different here. To correct for such motion in between successive volumes, a polynomial of 4-th degree is fitted to the different offsets within a volume (Fig. 3.14b). This results in a 2 dimensional function of axial displacement within one volume, which is corrected for by multiplying the axial Fourier spectra of each A-scan with a corresponding phase factor, according to the Fourier Shift theorem.

3.3.5 Segmentation

To analyze the neuronal function of the different cell types in the retina, the individual retinal layers must by segmented. For the following phase evaluation it is very important that the segmentation is robust, reproducible and as accurate as possible. At least the axial accuracy should be better than the resolution of the FF-SS-OCT system used.

A fundamental problem is that the OCT images are inherently noisy due to speckle formation. That is why, it is necessary to improve the image quality and to minimize the speckle noise in advance of the segmentation.

Another general problem for segmentation of OCT images of the retina is that the vascular system casts a shadow on the underlying layers, which causes interruptions or discontinuities in the individual retinal layers (Fig. 3.15 a). Furthermore, FF-SS-OCT has to deal with additional problems. For example, the photon yield is extremely low, due to the high imaging speed, which is why FF-SS-OCT has a lower signal-to-noise ratio (SNR) compared to other systems. In addition, FF-SS-OCT does not suppress multiple scattered photons, so that an additional noise source is present. For retinal layers, which are only separated by a few pixels, such as the inner segment/outer segment junction (IS/OS) or the OS tips and the RPE, the high noise level makes it hard to differentiate the respective layers. (Fig. 3.15 b). In addition, FF-SS-OCT images exhibit a poorer layer to layer contrast, which makes it hard to identify or segment layers with low intensity, such as the upper plexiform layer (Fig. 3.15 b).

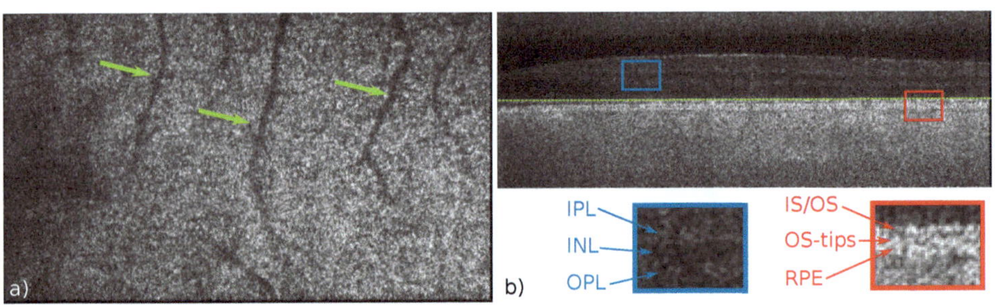

Figure 3.15: a) En-face image of a segmented IS/OS layer for a single measurement (layer is marked with green doted line in the B-scan of (b)). Shadows cast by the vessel system lead to gaps/discontinuities in the surface (marked with green arrows). b) B-scan of the same measurement. The blue display window illustrates the poor layer to layer contrast between IPL, INL, and OPL. The red display window illustrates the influence of noise for segmentation. Layers which lie close to each other like IS/OS, OS-tips, and RPE are difficult to distinguish in the presence of noise.

Moreover, for many clinical applications of OCT, it can be sufficient if only B-scans are segmented. However, for the the measurement of the functional connection between different neuronal layers, it is necessary to segment volumetric data sets, since connected neurons may be laterally shifted. Consequently, a segmentation algorithm should be chosen that can handle large data time efficiently. This can be achieved by reducing the complexity of the problem.

Additionally, some segmentation methods are susceptible to inhomogeneous illumination within the image, which causes fixed pattern noise. This can lead to incomplete or incorrectly segmentation. Therefore, an inhomogeneous intensity distribution in the collimated illumination beam of the FF-SS-OCT setup can lead to severe problems for the segmentation.

For clinical OCT diagnosis segmentation of the retinal layers is of utmost interest. For example, it is used to identify pathological changes in the individual layer thicknesses. So far, however, no algorithmic method has been able to prevail over other methods. This is mainly due to the fact that, depending on the OCT system used (TD, SD, SS, FF-TD and FF-SS), very different image qualities and imaging artifacts can be found, which can drastically influence the results of the segmentation. Nevertheless, there are several basic techniques for segmentation of OCT retinal images, all of which have their advantages and disadvantages and have been demonstrated in the past for different OCT systems and clinical applications. In the following only some of the most frequently used methods will be briefly introduced.

- **Pixel Classification Methods:** Pixel classification algorithms include algorithms like thresholding, classifier and cluster analysis. First attempts for OCT segmentation were done by 1D edge detection [159, 160] and thresholding approaches [161]. Usually, the contextual information of the sample structure is not taken into account and the segmentation is done individually for each A-scan, which makes this method susceptible to noise and an inhomogeneous illumination.

- **Pattern Recognition Techniques:** This technique divides its data sets into different categories. The algorithm seeks to find patterns or features that are common to all data sets within a category and that distinguish it from other categories. The different algorithms for this pattern recognition can be divided into supervised and unsupervised methods. Supervised techniques need training data. This means that a large number of training data sets have to be segmented manually. The algorithms use this training data as a reference for its own segmentation on a new data set. For OCT layer segmentation, especially support vector machine (SVM) algorithms were used as supervised pattern recognition techniques [162, 163]. In contrast to this, unsupervised pattern recognition algorithms do not need training data, but initial parameters are necessary for an efficient separation into the different categories. Unsupervised pattern recognition techniques used for retina segmentation are mainly cluster algorithms [164]. However, due to the poor layer contrast and high noise level, finding the right features for dividing the data into the different categories is difficult and in most cases, a pre-processing is necessary to better distinguish the categories or identify the features of interest.

- **Deformable Models:** Deformable models are algorithms that can identify boundaries between different regions. For this purpose, an initial, parameterized curve or surface is deformed under the influence of virtual internal and external forces to such an extent that it adapts to the boundary in the image data. Often a form of gradient information is used for this purpose, such as in the case of the gradient vector flow technique [165]. Deformable models are widely used in the segmentation of individual B-scans [166–168]. The disadvantage of this technique is that the curve must be initialized manually or by carefully chosen parameters. Furthermore, this technique can have difficulties with discontinuities in the curves, such as those caused by the vascular system in the retina.
- **Global Optimization and 3D Graphs:** In global optimization, segmentation is achieved by minimizing energy functions or maximizing probability functions. This usually includes 3D graph-based approaches that segment the OCT data by finding the shortest path or maximal flow through the graph designed on the image data [169]. A fundamental advantage of 3D graphs is, that weights describing both edge and regional information can be used. This in general increases the accuracy of the segmentation. In addition, there are also 3D graph-based approaches that cannot be classified as optimization, since they seek not to find the shortest path. These algorithms can be used to reduce the dimension of the data and/or for feature extraction. In combination with pattern recognition algorithms, they are hence suitable for segmentation [170–173].

For segmentation of FF-SS-OCT data, pixel classification methods are usually too susceptible to noise. Also, supervised pattern recognition and deformable models are unsuitable for this purpose, as they require manually segmented training data or initializations. Due to the especially poor layer contrast of FF-SS-OCT images, it is also difficult to identify the right feature of interest with an unsupervised pattern recognition method.

In order to segment the large, volumetric, highly noisy data set of the FF-SS-OCT images time-efficiently and accurately, a combination of a 3D graph with an unsupervised pattern recognition algorithm is used. The 3D graph method used here is called diffusion map [174–176]. This algorithm maps the data point depending on its connectivity into a new coordinate system: the diffusion distances. Therefore, not the shortest path is searched for, like in other graph-based algorithms, but all possible paths between two data points are summed up to determine their connectivity. For this reason, the algorithm is particularly robust against noisy data sets. By rearranging the data points according to their connectivity, the division of the OCT data into different categories becomes much easier. Therefore, a cluster analysis can be used to identify the different structures in the retina, after the data points are mapped into the diffusion distances. In the following, the mathematical basics of the diffusion map algorithm are presented as well as a short overview of different cluster algorithms.

3.3.5.1 Diffusion Maps

In order to map the data points into a new coordinate system, several steps are necessary. First, a kernel has to be created, which describes the local affinity of two data points i and j as the weights of a graph. Second, based on that graph a Markov chain is created, which maps the data into the new space. This mapping into a new coordinate system is realized in that manner, that data points belonging to the same structure are arranged close together, while data points from different structures are far away from each other. The thirds step is called spectral clustering and is done in order to reduce the dimensionality of the system. The three steps are explained in more detail in the following section. The fundamental mathematical thoughts are thereby analog to the description of Coifman et al. [174].

Construction of a Kernel With M being the set of data points, the kernel $k : M \times M \longrightarrow \mathbb{R}$ assigns every pair of data points $(i, j) \in M \times M$ a real valued weight, which should reflect the local connectivity of these two data points. Therefore, the kernel constitute the a priori definition of the local geometry and hence should be designed in a manner that it captures the specific feature of interest. Besides that, the kernel needs to satisfy the two following conditions:

$$k(i,j) = k(j,i) \tag{3.16}$$

$$k(i,j) \geq 0 \tag{3.17}$$

In the next step, the graph, that is constructed by the kernel, needs to be transferred into a Markov chain. A Markov chain is a sequence of stochastic events. The different positions in the sequence are often referred to as different time points. The event, i.e., the transition probability between two states, is independent of the path or sequence of the Markov chain and only depends on the current state. Commonly, these Markov processes are represented as a matrix, where the entry $p(i, j)$ of a Markov matrix P is the transition probability of getting from point i to point j within one step of the sequence or one time step. In order to create such a Markov matrix, the weight $k(i, j)$, which connects the two points i and j, needs to be converted into a transition

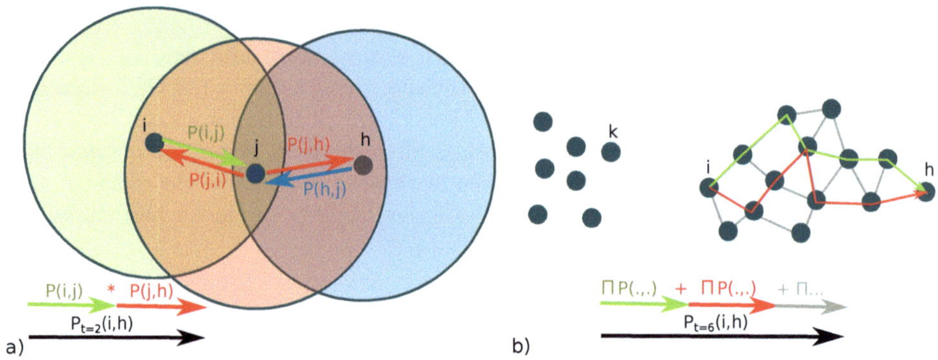

Figure 3.16: a) Illustration of the transition probability of a path with multiple time steps. The distance that can be covered within one time step are shown with the colored circles for the points i, j,and h. Point h cannot be reached from point i within one step. However, a connection can be established via another point (j) within two time steps. b) By using more time steps t, high transition probabilities can be reached for points far away from each other, as long as they are connected by path of length t (points i and h). The transition probability is calculated by the sum of possible paths of length t. However, if there is no path between two different points, an even greater time scale will still not connect them (points i and k).

probability $p(i,j)$ of these two points. Therefore, the previously generated weights need to be normalized. For this purpose, the sum of the weights within a row i of the kernel matrix are defined as row normalization factor $d(i)$ according to

$$d(i) = \sum_{j \in M} k(i,j). \tag{3.18}$$

Using this, the entries of the Markov chain can be calculated as

$$p(i,j) = \frac{k(i,j)}{d(i)}. \tag{3.19}$$

Mapping into a New Coordinate System The created Markov matrix gives a high transition probability between two data points if they are highly connected via their local geometry. In general, the local geometry can be defined as all kinds of features within the data such as similarity in the intensity or the Euclidean distance between two data points. For illustration purposes, it is assumed that the local geometry is defined by the distance between two points. This means that only transitions between data points close to each other are possible or likely. Yet, connections between distant points can be established over several time steps, if a path exists between them. For example, if point j has a high probability of transition to both point i and point h and vice versa (Fig. 3.16 a), then the probability of getting from point i to point h within two time-steps is high, even if the probability is zero for one time step. Therefore, if the Markov chain after t time steps is calculated, connections between distant positions of a structure are generated. Since the probability of reaching point h when starting at point i after t time-steps does not depend on the path or sequence of intermediate states in the Markov chain, all path of length t between these two points are summed up in the calculation (Fig. 3.16b). Points connected by many paths with a high probability, i.e., points belonging to the same structure, receive a high transition probability. In addition, individual 'wrong' paths over noise points lose significance, since they occur only rarely and become negligible when many paths are summed up.

To obtain the Markov chain P_t after t time steps, the t-th power of the matrix P is calculated. The matrix P_t can be seen as a mapping from the data set M to itself, i.e., an N-dimensional space, where N is the number of data points. To get an idea of the arrangement of the data points in the N-dimensional space, the diffusion distances $D_t(i,j)$ are introduced by

$$D_t(i,j) = \sqrt{\sum_{h \in M} |p_t(i,h) - p_t(j,h)|^2}. \tag{3.20}$$

If the kernel is chosen skillfully, then points within the same structure have high transition probability to each other and low transition probability to points from other structures. This means that data points i and j will have a similar transition probability to all points $h \in M$ if they belong to the same structure; that is, $D_t(i,j)$ will be very small. And vice versa, $D_t(i,j)$ becomes very large if i and j do not belong to the same structure. It follows, that data points belonging to the same structure are placed close to each other by the mapping, while points of different structures are far away from each other.

The problem that arises here is that clustering in N dimensions can become relatively complex. Furthermore, it can be assumed that most systems can be adequately described with much less than N dimensions. For this reason, the next step is a dimension reduction, also called spectral clustering.

Spectral Clustering The Markov chain P can be described completely by its eigenvalues λ and its eigenvectors Ψ by an eigendecomposition. For the previously created Markov chain, it can be shown that its largest eigenvalue is $\lambda_0 = 1$. The remaining eigenvalues can then be arranged in descending order of magnitude, correspondingly:

$$1 = \lambda_0 > |\lambda_1| \geq |\lambda_2| \geq \ldots \geq |\lambda_N| \tag{3.21}$$

The diffusion distances can be expressed as a function of its eigenvectors and eigenvalues by

$$D_t(i,j) = \sqrt{\sum_{h \in M} |p_t(i,h) - p_t(j,h)|^2} = \sqrt{\sum_{m=1}^{N} \lambda_m^{2t} (\Psi_m(i) - \Psi_m(j))^2}, \tag{3.22}$$

where Ψ_m are the eigenvectors belonging to the eigenvalues λ_m. The first advantage of the representation in terms of eigenvectors and their eigenvalues is that the t-th power of the Markov chain can now simply be carried out by raising the eigenvalues to the power of t, which makes the computation much more efficient. The second and much more important advantage is that the diffusion distance are weighted in each dimension with the corresponding eigenvalues. Since these are strictly smaller than 1, they approach 0 for $t \longrightarrow \infty$. Thus, if only the eigenvectors belonging to the largest eigenvalues are used for the calculation of the diffusion distance, only a negligible error occurs. For this reason, the dimension can be significantly reduced at this point without losing important information about the system.

3.3.5.2 Clustering

Clustering methods describe algorithms to find groups of similar objects in large data sets. A common representation of the objects is that of points in a multidimensional vector space, where each dimension describes a feature of that data set. Thereby, the task of the clustering algorithm is to identify accumulations of object points in the vector space as a group or cluster. Therefore, the mapping of the data points in the diffusion map is particularly suitable for clustering. There are basically three different methods of cluster algorithms, which can be better or worse suited for clustering depending on the arrangement of the object points in vector space or the requirements for clustering.

Partitioned methods Partitioned methods include cluster algorithms such as K-means [177], fuzzy C-means [178], and EM-algorithm [179]. Here, a certain number k of cluster centers is usually randomly selected from the objects and the remaining objects are then assigned to the cluster centers closest to them according to a chosen distance metric. If all objects are assigned to a cluster, the cluster centers are recalculated based on the objects within a cluster and all data points are reassigned to the new clusters. This is done until the position of the cluster centers no longer changes, or only changes within a predefined error.

Hierarchical methods Hierarchical methods create a hierarchy of clusters. On one side of the hierarchy, there is a single cluster that contains all objects, and on the other end of the hierarchy, there are N different clusters that each contain only one object,

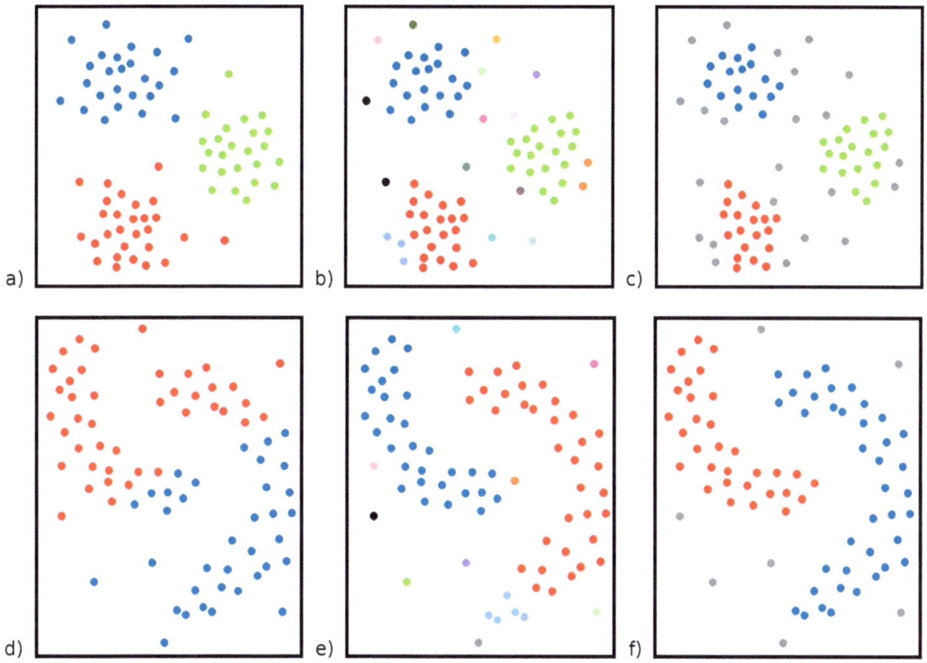

Figure 3.17: Comparison of different cluster algorithm on different data sets. a) and d) show the results of partitioned based cluster algorithm. b) and e) show the results of hierarchical based cluster algorithm. c) and f) show the results of density based cluster algorithm.

where N is the number of all objects. The clusters are subdivided in such a way that the data points within a cluster have smaller distances to each other than to the data points outside the cluster according to a previously selected distance function. As a result noise data are assigned to several own clusters (Fig. 3.17 b) and e)).

Density-based methods Density-based methods assume that the objects within a cluster are densely spaced and separated from other clusters by sparsely populated areas. Algorithms like DBSCAN [180] and OPTICS [181] work by identifying clusters via the density with which the data points are distributed within a cluster. Alternatively, there are also algorithms such as maximum margin clustering [182], which identify the sparsely populated regions between clusters in order to separate them. Thus, a cluster which has a large extension can still be identified as such if they are linked to each other via densely populated areas by density based methods (Fig. 3.17 c) and f)).

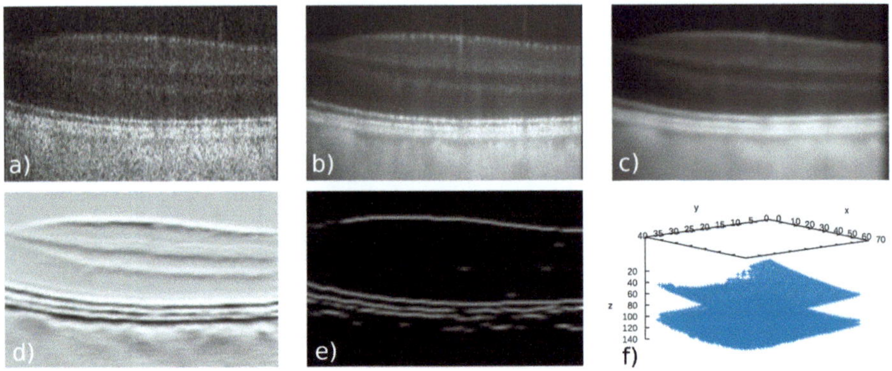

Figure 3.18: Demonstration of the Segmentation Steps with the same B-Scan a) B-scan from a single recorded volume near the fovea. b) B-scan from the incoherent average over all 70 volumes of one measurement. c) B-scan laterally coarse grained. d) B-scan after the axial Laplace filtering. e) Candidate points chosen for that B-scan. f) all candidate point chosen in the volume.

The best performing cluster algorithm depends on the structure of the data. An exemplary comparison of the algorithm presented here is shown in Fig. 3.17 for two different data set.

The advantage of partitioned cluster methods over hierarchical and density based cluster is that in each iteration the object points can be assigned to a different cluster. Misclustered objects can therefore be corrected in a more precise iteration. With hierarchical and density-based methods, once object points are assigned to one cluster, they cannot be reassigned to another in the next step. However, partition-based cluster methods assume that all clusters have the same extension and therefore will fail when there are clusters of very different sizes. Furthermore the number of clusters must be known in advance for partitioned based methods. With hierarchical clustering methods, on the other hand, there needs to be a criterion, which identifies the hierarchy level containing the desired clustering.

3.3.5.3 Algorithmic Implementation

In the following, the algorithmic implementation of the segmentation will be described. First of all, pre-processing is required to improve the image quality and to reduce the data size. Then the implementation of the diffusion map algorithm on the OCT data and the subsequent clustering of the data will be described.

Pre-Processing To reduce noise, all previously registered volumes of one measurement are incoherently averaged, i.e., only the modulus but not the phase was included in average (Fig. 3.18 b). This also reduces the data size, since only one averaged volume of each measurement needs to be segmented. However, a graph describing the whole volume would still roughly be of size $N \times N$ with $N \approx 640 \times 368 \times 256 \approx 52756480$. Solving an $N \times N$ matrix of this size would increase the calculation time enormously.

To further reduce the noise and complexity of the system, the resolution of the data is reduced. To preserve the axial information, this is only done in the lateral dimension. Therefore, the volume is subdivided in tiles of the size $10 \times 10 \times 1$ voxels. The average intensity of all voxels within one tile thereby describes the intensity of the new coarse voxel (Fig. 3.18 c).

For segmentation of the horizontal layers of the retina, an axial Laplace filter is performed by axially convolving the volume with $f_{\text{Lap}} = (1, -2, 1)^T$. Since the filter computes the second axial derivative of the volumes, it generates high values for horizontal edges (Fig. 3.18 d). To further reduce the data size only the four highest values for each A-scan of such a filtered volume are obtained as candidate points for further segmentation (Fig. 3.18 e and f). In general, these points are positioned at the NFL, the IS/OS , and the RPE. If the image quality is high it is possible to separate the OS tips from the RPE. In this case, the fourth point corresponds to the OS tips, while in the other cases they are randomly distributed over the volume. For a higher segmentation acuity, the axial positions of the candidate points are determined with sub-pixel precision via a parable fit of the maxima and their two axial neighbors.

In order to segment the plexiform layers, instead of a Laplace filter an axial step filter is computed by convolving the volume with $f_{\text{edge}} = (-1, -1, -1, 1, 2, 1, 2)^T$ in the axial dimension. The last four alternating digits are used to identify the layered structure of the IPL as precisely as possible. An edge filter creates high values for thicker layers like the IPL and OPL. Thus, the determined candidate points correspond to the plexiform layers and the RPE.

Only the chosen candidate points are then mapped by the diffusion map algorithm in the diffusion distances.

Construction of a Kernel The choice of the kernel k for the diffusion map is crucial for the outcome of the algorithm. Therefore, a kernel needs to be constructed, which creates high weights between two points of one surface and low weights between two points of different surfaces. Since the retinal layers can be described as more or less connected, a transition in a single time step is only possible if the points are close together. Furthermore, since the surfaces are horizontal, a lateral transition should be more likely than an axial transition. In order to consider these two aspects, a truncated, two-dimensional Gaussian-kernel in spherical coordinates $k(r, \theta)$ is used, which describes the weights as a function of the distance $r = \sqrt{(x_1 - x_2)^2 + (y_1 - y_2)^2 + (z_1 - z_2)^2}$ and

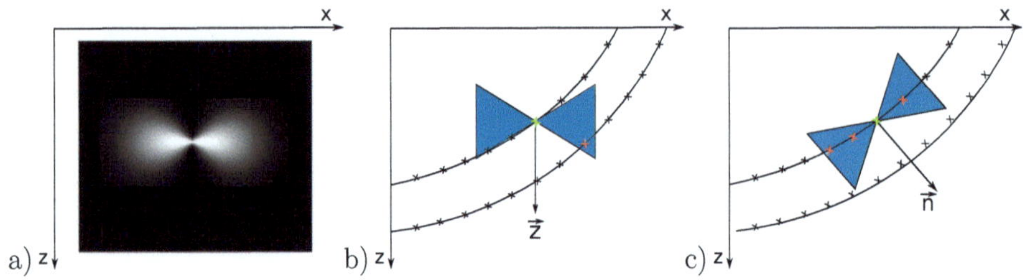

Figure 3.19: a) Side view of the kernel $k(i,j)$. b) Schematic side view of the kernel, illustrating how a curvature of two neighboring surfaces can lead to a false transition between the two surfaces. c) Schematic side view of the kernel, that is aligned according to the surface's curvature.

the angle $\theta = \arcsin\left(\frac{z_1 - z_2}{r}\right)$ relative to the z-plane of the two points $(x_1, y_1, z_1)^T$ and $(x_2, y_2, z_2)^T$. The kernel is thereby created according to

$$k'(r,\theta) = \begin{cases} \exp\left\{-\frac{r^2}{2\sigma_r^2} - \frac{\theta^2}{2\sigma_\theta^2}\right\} & \text{for } i \neq j, \\ 0 & \text{otherwise,} \end{cases} \tag{3.23}$$

with $\sigma_r = 1\,\text{px}$ and $\sigma_\theta = \frac{\pi}{7}$ (Fig. 3.19 a). However, with an axial resolution of 8 µm the IS/OS, the OS tips, and the RPE are only a few voxels apart from each other. Considering a small curvature of the surfaces a mutual transition of the different layers is possible using this kernel (Fig. 3.19 b). To specialize the kernel and to avoid such transition between different layers, a further parameter is added, which aligns the kernel according to the surface's curvature (Fig. 3.19 c). In order to do this the normal vectors of the surfaces are needed for every candidate point.

For an implicitly given surface $g(x,y,z) = 0$, the normal vector $\boldsymbol{n}$ can be calculated according to

$$\boldsymbol{n} = \begin{pmatrix} n_x \\ n_y \\ n_z \end{pmatrix} = \text{grad} g(x,y,z) = \begin{pmatrix} \frac{\partial}{\partial x} g(x,y,z) \\ \frac{\partial}{\partial y} g(x,y,z) \\ \frac{\partial}{\partial z} g(x,y,z) \end{pmatrix}. \tag{3.24}$$

With the intention of finding the surfaces, which are defined by the retinal layers in the volume $V(x,y,z)$, the axial Laplacian or an axial step filter was applied on the volume in the pre-processing. Because of that the surfaces are presented as maxima, at the respective maxima, the derivative in z will thus be zero. Thus the implicit surface can be described by

$$\frac{\partial}{\partial z} V'(x,y,z) = 0.$$

Here V' represents the axial filtered volume from the pre-processing. Hence the normal vector of the volume are calculated according to

$$\boldsymbol{n} = \begin{pmatrix} n_x \\ n_y \\ n_z \end{pmatrix} = \begin{pmatrix} \frac{\partial}{\partial x}\frac{\partial}{\partial z}V'(x,y,z) \\ \frac{\partial}{\partial y}\frac{\partial}{\partial z}V'(x,y,z) \\ \frac{\partial^2}{\partial z^2}V'(x,y,z) \end{pmatrix}. \tag{3.25}$$

Since the candidate points are determined with a sub-pixel resolution in the axial dimension, the normal vector has to be determined with a sub-pixel resolution as well. For this reason, the normal vectors of the two axial neighbors are calculated for each candidate point. The normal vector for the candidate point is then determined component-wise by a linear interpolation of those two normal vectors. In order to align the kernel according to the local curvature of the surface, the axis of rotation $\boldsymbol{a}$ and a rotation angle θ are needed. The axis of rotation is given by the cross product of the $\boldsymbol{n}$ with the z-axis,

$$\boldsymbol{a} = \boldsymbol{e}_z \times \boldsymbol{n} = \begin{pmatrix} -n_y \\ n_x \\ 0 \end{pmatrix}, \tag{3.26}$$

while the rotation angle is given by the angle between $\boldsymbol{n}$ and the z-axis

$$\theta = \arccos\left(\frac{|\boldsymbol{e}_z \cdot \boldsymbol{n}|}{|\boldsymbol{e}_z| \cdot |\boldsymbol{n}|}\right). \tag{3.27}$$

Since there will be different rotations for the positions i and j the weights $k'(i,j)$ and $k'(j,i)$ will not be the same. In order to still fulfill Equation (3.16) the weights are replaced with average of those two values according to

$$k(i,j) = k(j,i) = \frac{k'(i,j) + k'(j,i)}{2}. \tag{3.28}$$

Mapping in a New Coordinate System and Spectral Clustering The kernel is then, according to Equation (3.19), converted into a Markov matrix P. Due to storage capacity and computing efficiency reasons a sparse matrix is created. Therefore, all transition probabilities below a cut-off value of 1×10^{-2} are set to zero. This matrix is then decomposed into its eigenvalues and eigenvectors. Empirical analysis for FF-SS-OCT data showed that the highest four eigenvalues and the corresponding eigenvectors are sufficient to distinguish the different retinal layers in the diffusion distance from each other. Only after these four eigenvalues and eigenvectors are determined the Markov chain for greater time scales t is calculated, by raising the power of λ by t. The diffusion distances can then be described by:

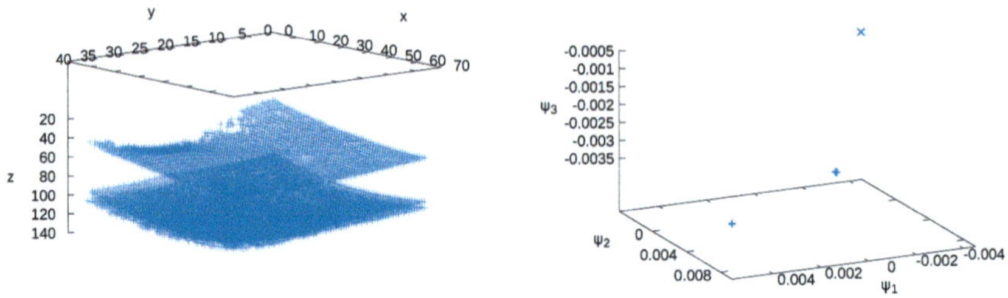

a) b)

Figure 3.20: Mapping into the diffusion coordinates: a) 3D view of all candidate points of one retinal volume. b) Mapping of those candidate points in the diffusion maps in dependency of the first three eigenvectors.

$$D_t(i,j) = \sqrt{\sum_{m=1}^{4} \lambda_m^{2t}(\Psi_m(i) - \Psi_m(j))^2} \tag{3.29}$$

The representation of the candidate points in their original form and to the basis of the diffusion distance are shown in Fig 3.20. To reach a stationary distribution of the Markov chain, the transition probabilities after 10.000 time steps are calculated.

Cluster Algorithm In order to choose an algorithm that is able to cluster the data in the diffusion map correctly and reproducible, one has to analyze how the data points are arranged in the diffusion distances and what requirements need to be considered for the searched clusters. For this particular system, there are three main problems or requirements.

One problem is that the image quality of different measurements can vary extremely depending on the subject or the position in the retina. In most cases the NFL, the IS/OS, and the RPE can be detected, in some cases additionally, the OS tips appear, and in the case of diseases like AMD it is not assured, that even the RPE can be distinguished from the IS/OS anymore. Since it, cannot be predicted how many clusters should be found by the algorithm, a stopping criterion is needed, that tells the algorithm when enough clusters have been found. A partition-based cluster method is therefore in general unsuited for this data. The second problem is that, in some cases, the spread of a cluster is larger than the distance between two clusters. This is especially the case for the clusters built by the RPE and the OS tips since these

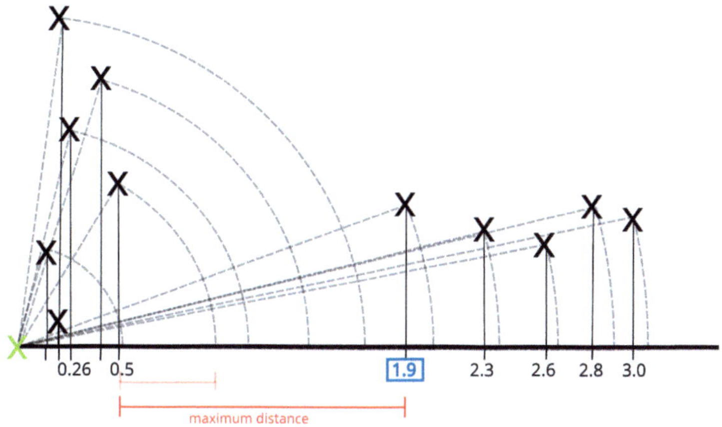

Figure 3.21: Schematic illustration for the calculation of the cut off distance: The Euclidean distance (gray) fails to identify the cut off distance between the two cluster correctly (largest discontinuity is marked with light red). The one dimensional distance (black) identify the distance between the two cluster correctly as largest discontinuity (red).

two layers are very close to each other. Since, even when using the specialized kernel an occasional transition between these layers cannot be avoided. For these data, a density-based method would be most suitable.

The third problem is that it is difficult to find a suitable cut-off distance or distance metric, which distinguishes the different clusters effectively from each other. It is particularly difficult to cluster the data set based on the density within one cluster since the density and thereby the intracluster distance can vary significantly between the different clusters. Furthermore, the density can also vary within a single cluster itself.

In order to address all of the above-mentioned problems and requirements a combination of a hierarchical and a density-based cluster method is used, which separates the different clusters not by their density but by identifying sparse areas in between different clusters.

Therefore a top-down hierarchy is used, i.e., for clustering, all data points are first considered and divided into two clusters according to a criterion defined later. Each newly created cluster is then further divided into two new clusters until the stopping criterion is met. The basic idea for the stopping criterion is that each ideal horizontal layer contains exactly $N_x \times N_y$ data points. In general, they will contain fewer points,

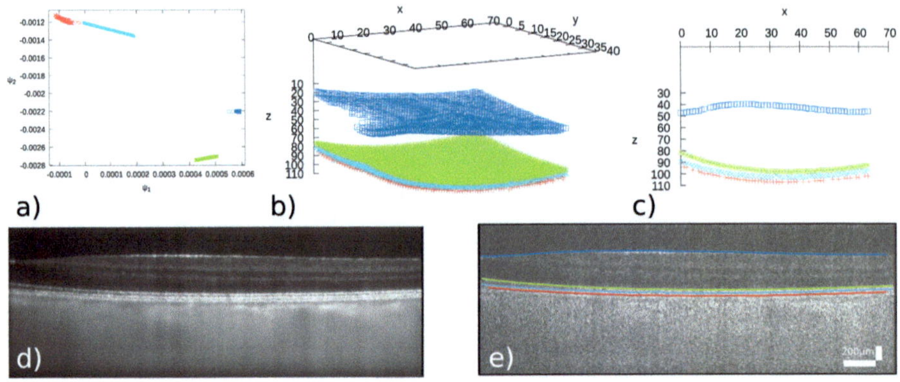

Figure 3.22: Results of the segmentation: a) Four different clusters in the diffusion coordinates, in dependency of the first two eigenvectors. b) 3D view of the clustered candidate points in the local space. c) Sectional image of the clustered candidate points. d) B-scan of the averaged volume. e) B-scan of the first volume of the measurement with the corresponding segmented surfaces. [183]

since vessels or noise lead to gaps in the surfaces, however, there will never be more data points. That is, if a cluster has fewer than $N_x \times N_y$ data points, it does not need to be further divided.

To subdivide the clusters, a cut off distance needs to be defined, that ensures that individual data points are not falsely separated from their clusters, while effectively separating different surfaces from each other. Since the density between different cluster and within single cluster can vary dramatically, an algorithm was created that identifies sparsely populated areas between two clusters. For this purpose a random point k is extracted from the currently considered cluster. For this random point the distance to all other points j of the cluster are calculated. In general there is always one dimension in the diffusion map in which two clusters are distinguishable particularly well. However, that does not necessarily mean that they can be distinguished in the other dimensions. Therefore no distance measure like the Euclidean norm is used here, because the directional information of the separation would get lost. This is illustrated in Figure 3.21, where the Euclidean distances are shown in gray and the distance in one dimension are shown in black. It becomes obvious, that the Euclidean distances would identify the sparsely populated area falsely. For this reason, not the Euclidean distances, but the distances for each dimension are calculated separately to maintain the direction dependence of the distances. These distances are then sorted by size for each dimension. Large discontinuities in the data indicate an empty space between different clusters. The larger distance that limits such discontinuities should therefore be suitable as a cut off distance to separate two cluster from each other. To identify this discontinuities, the sorted stacks are differentiated and the largest maximum of each derivative is determined. Thus, the dimension with the largest maximum or discontinuity is the dimension in which the data can be separated particularly well and in which, therefore, the clustering takes place. The corresponding cut off distance is determined as the larger distance corresponding to the maximum in the derivative. This is done for each iteration with a new chosen random point. Once, the data are completely clustered (Fig. 3.22) the data points can be mapped into its original coordinate system (Fig. 3.22 b and c). There, a polynomial of 4-th order is fitted via a least square fit to the data points of one cluster. This polynomial can then be used to align the surface of interest in a single depth via the Fourier-shift theorem.

3.3.6 Phase Evaluation

The previous processing steps ensure, that the phases of the backscattered light can be compared at exactly the same location of the speckle pattern during the time course of the measurement. As explained in Section 2.4.2 systematic changes of the scatterers can, therefore, be determined with nanometer precision by calculating the phase difference over time, since this referencing removes the random phase of the speckle pattern (Fig. 3.23 a).

Yet, biological tissue such as the retina has also intrinsic motion on a sub-cellular level. In addition to the systematic change of the scatterer, e.g., due to a stimulation of the retina, there is a constant chaotic or stochastic change of the scatterers due to metabolic processes or Brownian motion. While for a smaller time span this stochastic phase change is small and can be neglected (Fig. 3.23 a-c), on longer time scales those changes become more severe and eventually end up generating a completely new speckle pattern with a new random phase (Fig. 3.23 d-f). This means, that by using a simple referencing to one time-point, the phase differences will only yield meaningful results as long as the speckle patterns are still correlated. In order to evaluate the systematic phase changes beyond the decorrelation time of the speckle pattern, the stochastic phase needs to be minimized or separated from the systematic change. A similar problem exists in astronomic imaging, where the phase of the speckle pattern is altered due to turbulence in the atmosphere. Here, in the 1970s and 1980s algorithms were developed, for speckle interferometry, which enabled imaging through the turbulence in the atmosphere [184–188]. The mathematical description of the phase decorrelation in astronomic imaging is thereby the same as for phase-sensitive measurement in biological tissue. Therefore, these algorithms can be adapted for the purpose of evaluating phase changes in the retina beyond the decorrelation time [189]. One of these algorithms will be explained in the following. The following description corresponds to Spahr et al. [189].

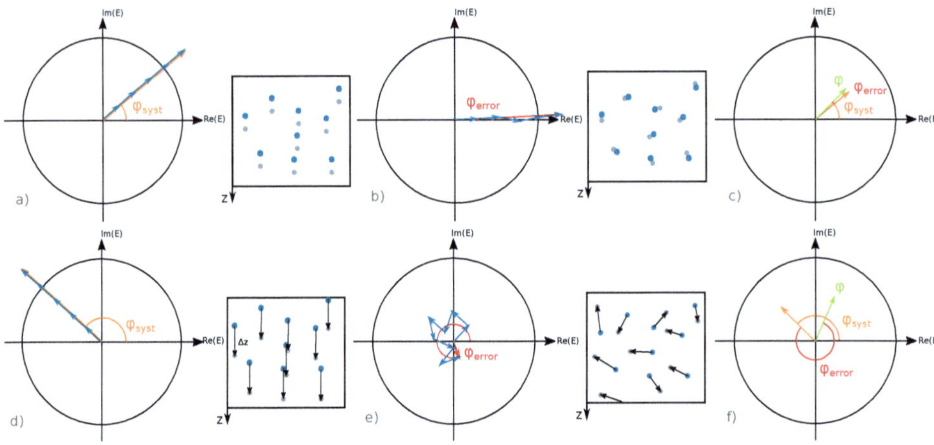

Figure 3.23: Schematic illustration of the speckle decorrelation: a) Short-time phase differences for the systematic phase changes. b) Short-time phase differences for the stochastic phase changes. c) Resulting phase φ for short time phase differences. The stochastic phase has only a small impact on the resulting phase. d) Long-time phase differences for the systematic phase changes. e) Long-time phase differences for the stochastic phase changes. f) Resulting phase φ for long time phase differences. The stochastic phase dominates the resulting phase.

In order to separate stochastic phase changes from the systematic phase changes the coherent acquired field $U(\boldsymbol{x}, t)$ can be described according to

$$U(\boldsymbol{x}, t) = \sum_j A_j(\boldsymbol{x}, t) e^{\mathrm{i}\Phi_j(\boldsymbol{x},t)} e^{\mathrm{i}\Phi(\boldsymbol{x},t)} = U_0(\boldsymbol{x}, t) H(\boldsymbol{x}, t), \tag{3.30}$$

where A_j describe the unknown amplitudes of each scatterer and $\boldsymbol{x}$ and t denote location and time, respectively. Here, $\sum_j A_j(\boldsymbol{x}, t)e^{\mathrm{i}\Phi_j(\boldsymbol{x},t)}$ represents the stochastic contribution on the phase change and will be denoted by $H(\boldsymbol{x}, t)$, while $e^{\mathrm{i}\Phi(\boldsymbol{x},t)}$ denotes the systematic contribution on the phase change, which will be denoted by $U_0(\boldsymbol{x}, t)$ in the following.

The phase change during a time step Δt can be described as the argument of

$$\begin{aligned} \Delta U(\boldsymbol{x}t, \Delta t) =& U(\boldsymbol{x}, t) U^*(\boldsymbol{x}, t + \Delta t) \\ =& \underbrace{U_0(\boldsymbol{x}, t) U_0^*(\boldsymbol{x}, t + \Delta t)}_{\Delta U_0(\boldsymbol{x},t,\Delta t)} \underbrace{H(\boldsymbol{x}, t) H^*(\boldsymbol{x}, t + \Delta t)}_{\Delta H(\boldsymbol{x},t,\Delta t)}. \end{aligned} \tag{3.31}$$

Here, the first term ΔU_0 is determined by the systematic motion of the scatterer and ΔH by the stochastic motion.The influence to the phase change of the coherent field by the stochastic phase can be described according to Equation 3.30by:

$$\begin{aligned}\Delta H(\boldsymbol{x},t,\Delta t) &= \left(\sum_j A_j(t)e^{\mathrm{i}\Phi_j(\boldsymbol{x},t)}\right)\left(\sum_k A_k(t+\Delta t)e^{\mathrm{i}\Phi_k(\boldsymbol{x},t+\Delta t)}\right)^* \\ &= \left(\sum_j A_j(t)A_j^*(t+\Delta t)e^{\mathrm{i}(\Phi_j(\boldsymbol{x},t)-\Phi_j(\boldsymbol{x},t+\Delta t))}\right) \\ &+ \left(\sum_{j,k,j\neq k} A_j(t)A_k^*(t+\Delta t)e^{\mathrm{i}(\Phi_j(\boldsymbol{x},t)-\Phi_k(\boldsymbol{x},t+\Delta t))}\right)\end{aligned}$$

Here, Φ_j and Φ_k are statistically independent. The second term determines the phase difference of a scatterer j to a different scatterer k after a time Δt. Since Φ_j and Φ_k are statistical independent, this means that the second term describes a random walk in the complex plane leading to a small magnitude and a random phase. The first term, on the other hand, determines the phase difference of a scatterer to itself after a time Δt. For small time differences Δt phase changes of a scatterer to itself will be small and accordingly, the phases of the first term will all be close to zero and the magnitudes will all add up constructively. Given a sufficient number of values from different scatterers, the contribution of the second term is thus small compared to the first term and only leads to an insignificant phase error. For this reason, it can be assumed that for short time differences the stochastic phase will be close to zero and the phase difference is therefore dominated by the systematic phase (Fig. 3.23 a-b). However, using short time phase referencing followed by a summation of all calculated phase differences will lead mathematically to the same results as long time phase referencing, since the phase noise will accumulated over time. This can be seen from Figure 3.24 b and e. Here, a cross-referencing matrix $C_{t,\Delta t} = \Delta U(\boldsymbol{x},t,\Delta t)$ is shown, which contains all possible temporal referencing between t and Δt.

The result of a single event, i. e., the value of a single phase error, cannot be predicted and therefore there is no way to remove this single phase error per se. However, since the phase errors for short time spans are independently distributed around zero, the ensemble average of N independently phase errors will approach zero, if N is chosen sufficiently large.

Thereby, the influence of the phase error is minimized, so that for a sufficiently large ensemble the phase of the stochastic contribution can assumed to be zero. The systematic phase difference however is constant for each measurement and the ensemble average will therefore be the same as for a single measurement.

By evaluating the phase changes only for short time differences, which are within the correlation time of the speckles, and averaging them over a large ensemble, the influence of the stochastic phase can be removed and only the systematic phase remains (Fig. 3.24 c and f). Summing the phases afterwards yields the time course of the systematic changes in the retina. The ensemble average of short time differences can be described by

$$\begin{aligned}\langle \Delta U(\boldsymbol{x}, t, \Delta t)\rangle =&\langle U(\boldsymbol{x}, t) U^*(\boldsymbol{x}, t+\Delta t)\rangle \\ =&\langle \Delta H(\boldsymbol{x}, t, \Delta t)\rangle\langle \Delta U_0(\boldsymbol{x}, t, \Delta t)\rangle.\end{aligned} \tag{3.32}$$

$\langle \Delta U(\boldsymbol{x}, t, \Delta t)\rangle$ is similar to the time-autocorrelation of U, however, the averaging is done over the ensemble instead of the time t and thus it is a function of Δt as well as of t. For that reason, the cross referencing matrix maintains comparably large magnitudes as long as $U(\boldsymbol{x}, t)$ and $U(\boldsymbol{x}, t+\Delta t)$ are in good correlation to each other. Thus, if the volumes start to decorrelate not only a non-negligible phase error occurs but also the magnitude becomes smaller. Therefore, the magnitude of the cross-referencing matrix can be used as weights indicating the degree of correlation between two volumes.

If, for example, a single volume in the measurement is altered by a saccadic motion, there will be no correlation in the speckle pattern even for small time differences. Therefore, a phase error is introduced if this volume is used as a reference volume, which remains for the rest of the time course. The influence of such a single event can be reduced by taking multiple ways with different Δt for stepping through the cross-referencing matrix $\langle \Delta U(\boldsymbol{x}, t, \Delta t)\rangle$. If additionally each time-step is weighted by its corresponding magnitude such an outlier has nearly no influence on the resulting time course anymore.

3.3.6.1 Algorithmic Implementation

The cross-referencing matrix is calculated from the series of OCT volumes $U(\boldsymbol{x}, t)$ according to

$$\Delta U(\boldsymbol{x}, t, \Delta t) = U(\boldsymbol{x}, t) U^*(\boldsymbol{x}, t+\Delta t).$$

Here $U(\boldsymbol{x}, t)$ is the complex OCT signal at position $\boldsymbol{x}$ and time t.

For functional analysis of the different neurons, the change of thickness of the different neuronal layers is of interest. Therefore, the phase difference between two different retinal layers z_1 and z_2 is determined. To improve the SNR for the phase evaluation each layer is in general axially averaged over several pixels according to:

$$\Delta U^{z_1-z_2}(x, y; t, \Delta t) = \left(\sum_{z\in Z_1} \Delta U(x, y, z; t, \Delta t)\right)\left(\sum_{z\in Z_2} \Delta U(x, y, z; t, \Delta t)\right)^*,$$

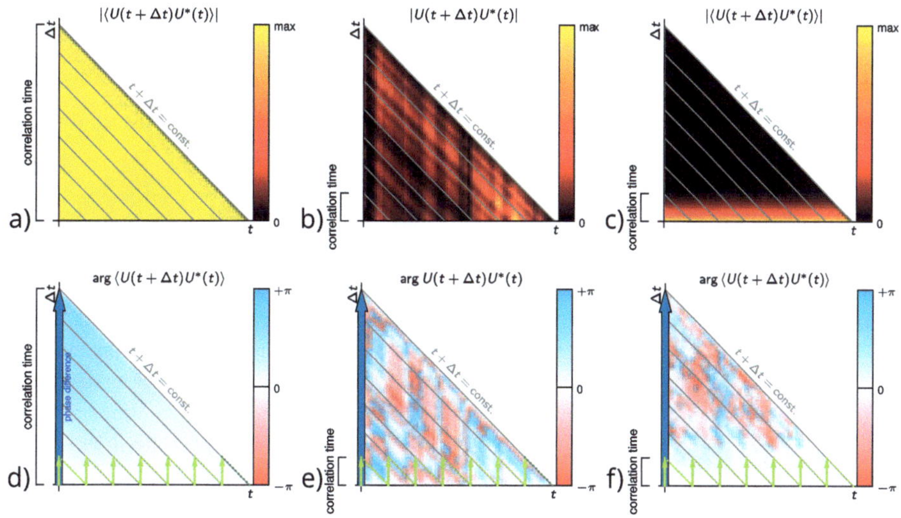

Figure 3.24: a,d) Exemplary cross-referencing matrix with magnitude and phase assuming infinite correlation time. Total phase changes can either be computed directly (blue arrow) or using path consisting of many short time references (green arrows). The result does not depend on the path. b,e) Magnitude and phase of the cross-referencing matrix without ensemble averaging of a real measurement; The result is also path independent but the phase cannot be extracted beyond the correlation time. c,f) Magnitude and phase of ensemble averaged cross-referencing matrix. Direct phase differences (blue) cannot be used for phase extraction, but using short time references (green) method can be applied.

where Z_1 and Z_2 indicate the depth pixels used for averaging.

In the next step, a sufficiently large ensemble of the measurements under the same condition needs to be generated, which means several independent measurements at the same position and the same time need to be generated. This cannot be realized for functional retinal imaging. However, since the stochastic phase changes of adjacent speckles are independent, whereas the systematic change is assumed to be constant in the closer environment, multiple lateral pixels will form an ensemble with the same properties. As some areas of the retina are stimulated for functional analysis and others are not, it is not possible to determine in advance which areas show identical systematic phase changes. Therefore, only phases that are close to each other are used for the ensemble average at this point, by convolving the phases with a lateral Gaussian function according to:

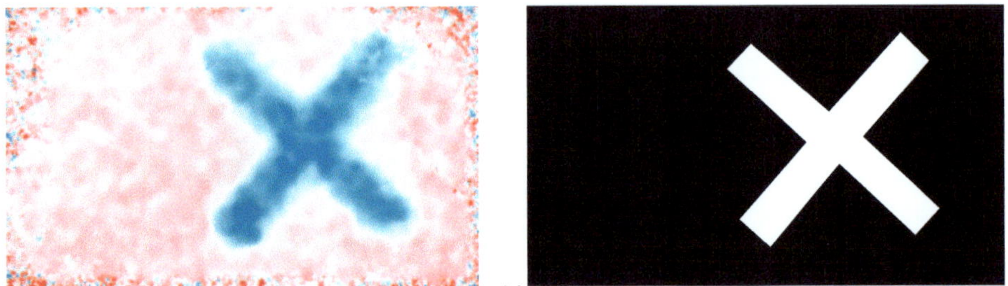

a) b)

Figure 3.25: a) En-face image of the phase difference between the IS/OS and the OS-tips after 3 sec stimulation with an x-shaped pattern. For the ensemble average the phases were convoluted with a Gaussian filter. b) Corresponding binary mask. The mask covers the area, where the functional/systematic changes arise. Thereby the ensemble average can be significantly improved.

$$\left\langle \Delta U^{z_1-z_2}(x,y;t,\Delta t)\right\rangle \approx G_{\sigma^2}(x,y) * \Delta U^{z_1-z_2}(x,y;t,\Delta t)$$

Here, $*$ denotes a circular convolution and $G_{\sigma^2}(x,y)$ is the Gaussian function with $\sigma = 4$ pixel. By doing so, an en-face image of the phase changes in the retina is generated (see Fig. 3.25 a). From this en-face image, the area in which a functional change occurs can be determined. Therefore, the area can be averaged using an adapted mask (see Fig. 3.25 b) to obtain a more stable ensemble average according to

$$\left\langle \Delta U^{z_1-z_2}(t,\Delta t)\right\rangle \approx \frac{1}{|M|}\sum_{(x,y)\in M} \Delta U^{z_1-z_2}(x,y;t,\Delta t).$$

Here, $|M|$ is the number of pixels within the mask M. The weights, which were used to identify the reliability of the values of the cross-referencing matrix, were calculated by

$$w(t,\Delta t) = T\left(\left[G_{\sigma^2}(x,y) * \frac{\Delta U^{z_1-z_2}(x,y;t,\Delta t)}{|\Delta U^{z_1-z_2}(x,y;t,\Delta t)|}\right] - b\right) \tag{3.33}$$

for the en-face images and

$$w(t,\Delta t) = T\left(\sum_{(x,y)\in M} \frac{1}{|M|}\frac{\Delta U^{z_1-z_2}(x,y;t,\Delta t)}{|\Delta U^{z_1-z_2}(x,y;t,\Delta t)|} - b\right) \tag{3.34}$$

for the curve extracted by the mask M. Here, T is ramp-function to cut-off negative values:

$$T(x) = \begin{cases} x & \text{for } x \geq 0 \\ 0 & \text{for } x < 0 \end{cases}$$

The truncation of negative values and the introduction of b are supposed to keep the weight for randomly distributed ΔU at 0 [189].

If the observed systematic changes are larger than $\frac{\lambda_0}{4}$, then the phases will wrap. To obtain a continuous time course, the phase of the cross-referencing matrix has to be unwrapped along the Δt axis.

After the phase is unwrapped, the actual time course of the systematic change in the retina can be obtained. The system created here is similar to the over determined system in Equation (3.7) used for the registration. Here, the vector indices again correspond to different times, and the system is given by

$$\Delta\mathbf{\Phi} = \begin{pmatrix} \Delta\Phi_{0,1} \\ \Delta\Phi_{0,2} \\ \vdots \\ \Delta\Phi_{1,2} \\ \vdots \end{pmatrix} = \begin{pmatrix} 1 & -1 & 0 & \cdots & 0 \\ 1 & 0 & -1 & \cdots & 0 \\ & & \vdots & \ddots & \vdots \\ 0 & 1 & -1 & & 0 \\ & & \vdots & & \vdots \end{pmatrix} \begin{pmatrix} \Phi'_0 \\ \Phi'_1 \\ \vdots \\ \Phi_{N-1} \end{pmatrix} = D\mathbf{\Phi}',$$

and has no unique solution, since for different initial values the time course $\mathbf{\Phi}'$ will be different as well. The determination of the time course $\mathbf{\Phi}'$ is thereby done similar to 3.10 with the formulation of a least square problem:

$$\|W(D\mathbf{\Phi}' - \Delta\mathbf{\Phi})\|^2 + \mu_1\|\Gamma_1\mathbf{\Phi}'\|^2 + \mu_2\|\Gamma_2\mathbf{\Phi}'\|^2 + \mu_3\|\Gamma_3\mathbf{\Phi}'\|^2 \longrightarrow \min_{\mathbf{\Phi}'}$$

with three regularization terms $\mu_1\|\Gamma_1\mathbf{\Phi}'\|^2, \mu_2\|\Gamma_2\mathbf{\Phi}'\|$ and $\mu_3\|\Gamma_3\mathbf{\Phi}'\|$. Here, $W = \mathrm{diag}(w_0, w_1, \cdots, w_{N-1})$ is a diagonal matrix containing the weights computed by Equations (3.33) and (3.34). The pseudo inverse of this system is again obtained according to

$$\Phi' = \left(D^T W^2 D + \mu_1^2 \Gamma_1^T \Gamma_1 + \mu_2^2 \Gamma_2^T \Gamma_2 + \mu_3^2 \Gamma_3^T \Gamma_3\right)^{-1} D^T W^2 \Delta\mathbf{\Phi}.$$

For the regularization terms

$$\Gamma_{1,ij} = \begin{cases} 1 & \text{for } i = j = 4 \\ 0 & \text{otherwise} \end{cases},$$

$$\Gamma_{2,ij} = \begin{cases} 1 & \text{for } i = j \neq 0 \\ -1 & \text{for } i = j + 1 \\ 0 & \text{otherwise} \end{cases},$$

$$\Gamma_{3,ij} = \begin{cases} 2 & \text{for}\, i = j, i \neq 0\, \text{and}\, \text{i} \neq \text{N} - 1 \\ -1 & \text{for}\, i = j + 1\, \text{or}\, i = j - 1 \\ 0 & \text{otherwise} \end{cases},$$

are chosen with $\mu_1 = 0.09, \mu_2 = 0.02$, and $\mu_3 = 0.05$. These matrices keep the value of 4-th volume, which is the reference volume before the stimulation starts, as small as possible and in general keep the first and second derivative of the phase time-course also small. Without those regularization terms the matrix D is not invertible if there are gaps in the data.

4 Results

By using FF-SS-OCT we were able, in 2016, to demonstrate for the first time phase-sensitive functional measurement in the human retina [190]. Back then, we showed that an increase in the optical path length of the photoreceptor OS occurs when they are stimulated with light.

In the scope of this thesis, improved post-processing is used to systematically characterize this effect and to perform functional phase-sensitive measurements in the upper neuronal retinal layer.

Compared to the work shown in 2016, the registration has become more stable and now takes rotations and deformations into account, which allows a phase-stable evaluation of measurements taken over a longer period of time. The segmentation has changed from a pixel classification algorithm to a graph-based algorithm, which allows more accurate and reproducible identification of the structures of interest. The phase evaluation has changed from a simple time referencing of the phase to a statistical evaluation of the phase, which allows evaluation over times exceeding the decorrelation time of the speckle pattern. As a result, more robust, more sensitive, and longer phase evaluations are now possible than 4 years ago.

Using these improvements, the biological relevance of the measured functional changes of the phase was investigated. On the one hand, this work moves to significantly lower stimulation energies, which are close to the mesopic range of vision, in order to distinguish the different contributions of rods and cones. On the other hand, it is shown that functional changes can additionally be observed in the IPL. For a biological classification of the functional changes in both layers, i.e., in the photoreceptor layer as well as in the IPL, are investigated with respect to their spatial and temporal characteristics.

Since these measurements and the data handling are still extremely complex and time-consuming, it was not possible to measure the signals systematically in multiple volunteers. Therefore, this section presents systematic measurements in a single volunteer. Yet, functional measurements were reproduced in several individuals.

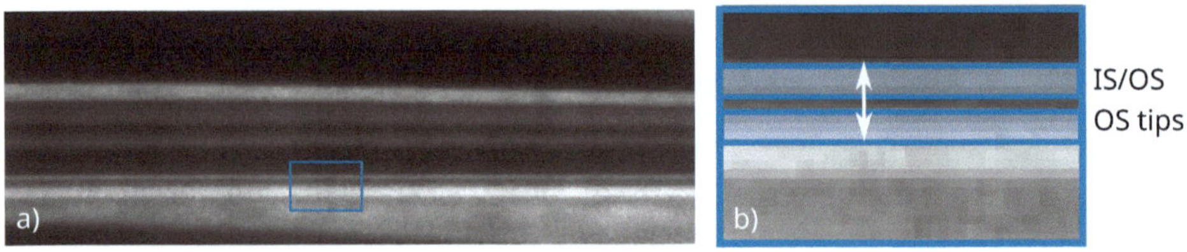

Figure 4.1: Depth used for the phase evaluation of the photoreceptor OS. a) B-scan of the subjects retina in 3 mm periphery. b) Close-up of the blue marked region in (a) the blue marked layers correspond to the layers used for the phase evaluation

4.1 Intrinsic Optical Signal in the Photoreceptor Cells

For several reasons, changes in the photoreceptor cells are the natural first choice when investigating functional neuronal changes in the retina. First, the photoreceptors are the cells that start the visual process and therefore are directly activated by the stimulation pattern. This means that the functional activity of the photoreceptors should appear exactly in the illuminated area of the retina and should therefore be easy to identify. In contrast to this, the other neurons in the retina are not directly activated by the light stimulus but by the preceding neuronal layers. Second, for a good SNR of the phase difference, it is important to evaluate the phases of layers that have a high intensity in the OCT image. The inner segment-outer segment junction (IS/OS) produces a very bright, sharply isolated layer in the OCT image and is therefore well suited for phase evaluation.

Thus, for the functional evaluation of the photoreceptors, the IS/OS is segmented and the volumes are aligned along this line. To further improve the SNR, the phase information of the IS/OS is averaged over three depth pixels. As reference layer, the tips of the OS are chosen. For the phase evaluation, it is assumed that these are parallel to the IS/OS. If not mentioned otherwise the phase difference was measured to a layer five depth pixels below the IS/OS and again averaged over three depth pixel (see Fig. 4.1). Since the IS/OS represents the upper end of the photoreceptor OS and the OS tips the lower end, such a phase referencing measures the changes of the photoreceptor OS's optical path length.

4.1.1 Signal Analysis

4.1.1.1 General Observation

For the functional investigation of the photoreceptor OS, the retina was stimulated with an x-shaped pattern in 3 mm periphery (Fig. 4.2 a). The stimulation was continuous over the whole measurement duration of 8 s and had an intensity of 57 nW/mm^2. This intensity was chosen in order to expose the subject to a clearly recognizable stimulus without producing an afterimage, and thus preventing complete bleaching of the stimulated area. Figure 4.2 b shows an en-face image of the optical path length change in the photoreceptor OS after 8 s of stimulation. It can be seen that the stimulation leads to an increase in the optical path length over time. Thereby the area where these changes arise is sharply limited to the stimulated area in the retina.

For the quantitative evaluation of the functional changes, the signals were averaged over the stimulated area (Fig. 4.2 b outlined region). Figure 4.2 c shows the corresponding time course of the optical path length changes in this stimulated area. After an initial rapid increase of the optical path length with approx. 100 nm/s, the curve flattens after about 2 s and reaches a final maximum state of approx. 125 nm.

The measured changes are significantly larger than the phase noise, so that generally a good SNR of the functional signals in the photoreceptor OS can be achieved. To ensure that quantitative information about duration and signal intensity can be drawn from the measured curves, the reproducibility of the signal were first evaluated.

4.1.1.2 Reproducibility

To investigate the reproducibility of the functional signal, five different measurements were taken with a stimulation of 57 nW/mm^2 over 8 s on three different days, partly several month apart. The time courses of the functional signal of this measurements are shown in Figure 4.3 a in gray. The blue curve corresponds to the mean value of these measurements. The time course in the various measurement is generally well reproducible, even over long times.

After the first 125 ms of stimulation the different time-courses have a maximum deviation of 2.7 nm. The initial slope, which can be extracted by a linear approximation of the optical path length changes after 125 ms, varies accordingly between 92.6 nm/s and 114.4 nm/s. However, the deviation becomes larger towards the end of the measurement, achieving a maximum deviation of 22.5 nm after 8 s stimulation.

This deviation could be caused by an inaccuracy of the phase detection. For example, due to a residual influence of the unsystematic phase change which cannot be neglected. Another possible influence for this increasing deviation during the measurement could be due to eye movements. The FF-SS-OCT system used here is not able to compensate the stimulation according to the eye movements As a result, movements or drifts of the eye during the measurement shift the stimulation over the retina with

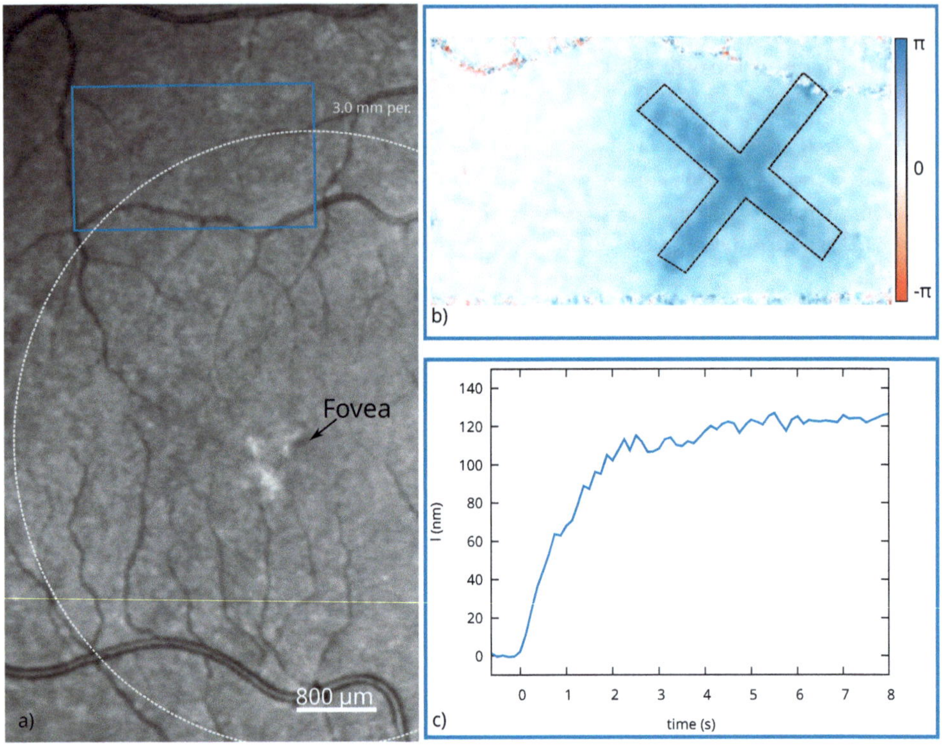

Figure 4.2: Result of the phase evaluation: a) SLO image of the volunteer retina obtained with a Spectralis OCT from Heidelberg Engineering. The field of view which were imaged is marked in blue. b) En-face image of the optical path length changes changes in the PR OS measured after 8 s continuous stimulation with 57 nW/mm^2. c) Time course of the measured optical path length changes.

time and thus increase the stimulated area. If there is a lot of movement during a measurement, the total number of photons stimulating the retina is distributed over a larger area or more photoreceptors, which results in less stimulation of each individual photoreceptors. In order to evaluate the influence of the eye movements on the time-course of the functional signal, the maximum offset from the initial position as determined by the registration within each measurement was compared with the optical path length change after 8 s. The results are plotted in Figure 4.3 b. Volumes destroyed by saccades were not taken into account because their actual offset cannot be determined correctly.

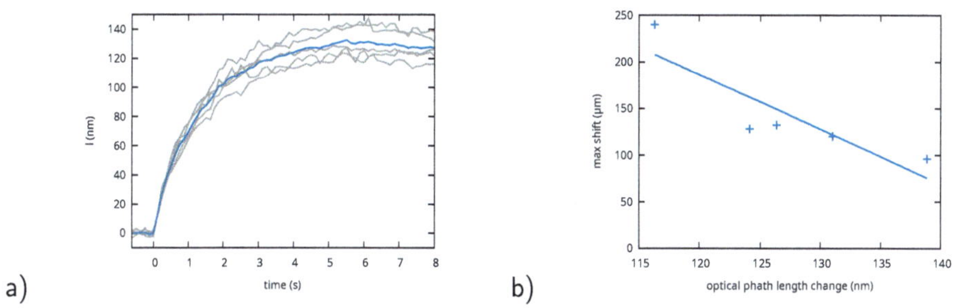

Figure 4.3: a) Time course of the OPL change in the photoreceptor OS. The gray curves are individual measurement under the same conditions (8 s stimulation, with an intensity of 57 nW/mm^2 in 3 mm periphery). Blue curve is the average time course. b) Maximum shift during the measurement, due to eye movements versus the measured optical path length changes after 8 s stimulation.

In general, a smaller functional signal is correlated with larger movements or drifts during the measurement. Thus, it can be assumed that the degree of eye motion has an influence on the resulting signal strength.

4.1.1.3 Functional Imaging of Single Cones

The phase evaluation described in Section 3.3.6 averages the phase over a sufficiently large lateral area to minimize the stochastic phase error. In most evaluations, this is done by a lateral Gaussian filter with a standard deviation of 4 px. Nevertheless, this lateral averaging limits the lateral resolution with which the phase signal can be assigned to certain morphological structures.

In the periphery of the retina, the diameter of the cones is approx. 3 µm, which corresponds to only a few pixels in the FF-SS-OCT volumes. In order to investigate whether the functional change can be assigned to individual photoreceptors or whether it is a global effect, a pixel-accurate phase evaluation is desirable.

In order to achieve a better lateral resolution of the phase evaluation, in this section, the lateral phases were not averaged. Due to the missing Gaussian filter, the noise of the phase evaluation was significantly higher and degraded the phases, especially for longer measurements, since the speckle patterns decorrelate over time. For shorter measurement duration, where the noise is still relatively low, the functional signal is also still very weak. Therefore, a simple time reference of the phase, without lateral averaging, did not achieve a good SNR as well.

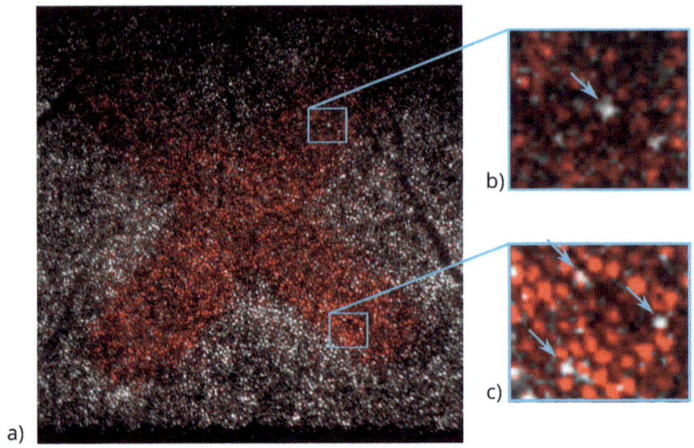

Figure 4.4: Functional phase change in the photoreceptor OS superimposed with the scattering intensity of the photoreceptor layer. a) En-face image of the functional change after 8 s of stimulation with 1.9 µW/mm^2 superimposed with the scattering intensity of the photoreceptor layer.

In order to achieve a good SNR at maximum resolution, the phases were averaged temporally instead of spatially. For this purpose, the cross-referencing matrix of Section 3.3.6 is filtered with a Gaussian Kernel in the t and delta Δt axes. By using a temporal Gaussian filter instead of a spatial Gaussian filtering, the noise is minimized while the functional phase change is preserved and a higher lateral resolution is reached. However, averaging the phase over time alters the time course and is particularly erroneous in regions where the functional phase changes rapidly. Therefore, after temporal averaging the dynamic of the functional change were not measurable with good accuracy, but qualitative statements about the functional reaction of individual pixels were possible. In order to obtain the highest possible response of the photoreceptor cells, a stimulation power of 1.9 µW/mm^2 over 8 s was used for measuring. Figure 4.4 shows a corresponding en-face image of this phase evaluation after 8 s stimulation superimposed with the corresponding intensity image of the evaluated plane.

Again, the stimulated area on the retina can be clearly identified by the functional changes of the phases without lateral averaging, but within the stimulated area also single cones can be observed, which do not contribute to the functional changes in the phase (see Fig. 4.4).

Both technical and physiological causes for such a lack of functional signal in individual photoreceptors are possible. On the one hand, such an absence of the functional signal for individual photoreceptors can originate if the phase from one of the two reference layers of that particular photoreceptor cannot be determined with a sufficiently large SNR, i.e., either the IS/OS or the OS tips lack sufficient signal. On the other hand, this can also be caused by the functional change, i.e. the physiological activity of the photoreceptor, not being large enough to be detected.

4.1.1.4 Dependence on the Stimulation Strength

In a next step, the dependence of the functional signal on the stimulus intensity was investigated. For this purpose, measurements in 3 mm periphery were performed and again the retina were continuously stimulated by an x-shaped pattern. To vary the stimulation irradiances between 1.92 µW/mm^2 and 3 nW/mm^2, different OD-filters were introduced into the stimulation arm. The used OD-filters and the corresponding irradiances on the retina are listed in Table 4.1. For comparison with published work and better relation to physiological effects, photon flux, luminance, and degree of photoreceptor bleaching were calculated. The conversion into photometric unit were done according to Equations 2.19 and 2.20. Based on the spectrum $\Phi_e(\lambda)$ of the LED from Section 3.1 and the luminous efficiency function $V(\lambda)$ [99], the integral $\int \mathrm{d}\lambda\, V(\lambda)\Phi_e(\lambda)$ gives an efficiency factor for the stimulation of 0.61. For the conversion into luminance, a pupil area of $A_p = 50.26\,\mathrm{mm}^2$ is used for a maximum pupil diameter of 8 mm and the distance from cornea to retina $D = 17\,\mathrm{mm}$ is assumed.

The converted values for the different stimuli are shown in Table 4.1 as well as the number of photons per µm^2 and the percentage of photopigments bleached within one second of stimulation in rods. Again, a stimulation efficiency factor of 0.61 was assumed, a base area for rods of $A_{\mathrm{rod}} = 3.14\,\mu\mathrm{m}^2$, and an absolute photopigment number per photoreceptor of $N_{\mathrm{rod}} = 10^8$ [103].

In Figure 4.5 the different equivalent stimulation luminances are compared with the luminances which prevail in photopic and scotopic vision [191]. However, it must be noted that in the measurements of photopic and scotopic vision, the pupil size has adapted to the prevailing light conditions. This means that for high irradiances the pupil diameter decreases to approximately 2.5 mm, in order to reduce the incidence light on the retina. In the measurements performed here, however, a blocking of the pupil reflex ensures that the pupil diameter is constant at 8 mm. While this makes little difference for the lower luminances, where the pupil diameter is still relatively large, the highest luminance (4600 cd/m^2) would have to be multiplied by a factor of 10 to be comparable to the measurements of [191].

The highest used luminances falls in the range of photopic vision, where only cones are active, while the lower luminances are in the upper mesopic region, the transition region from photopic to scotopic vision, where both rods and cones are active.

Additionally, the luminance caused by the broad sweeper of the OCT imaging was calculated to determine the influence of retinal stimulation caused by the measurement itself. For this, an intensity of 0.83 mW/mm^2 is used. Since for the central wavelength $\lambda_0 = 840\,\mathrm{nm}$, $V(\lambda)$ is not defined anymore, the closest available wavelength (830 nm)

OD-filter	Irradiance $[\frac{\mu W}{mm^2}]$	$\frac{photons}{\mu m^2 \times s}[\times 10^6]$	luminance $[\frac{cd}{m^2}]$	PPB R $[\frac{\%}{s}]$	PPB C $[\frac{\%}{s}]$
0.0	1.9	5.3	4600	10.0	40.0
0.6	0.52	1.44	1240	3.2	12.8
1.0	0.19	0.53	460	1.0	4.0
1.3	96.0×10^{-3}	0.27	230	0.5	2.0
1.6	57.0×10^{-3}	0.16	137	0.3	1.2
2.0	15.4×10^{-3}	0.04	37	0.08	0.32
2.3	7.0×10^{-3}	0.02	16.8	0.04	0.16
2.6	3.0×10^{-3}	0.008	7	0.02	0.08

Table 4.1: List of the used stimulation intensities expressed in intensity, luminance, photon flux and the percentage of photopigment bleaching per second (PPB R) in rods and (PPB C) in cones

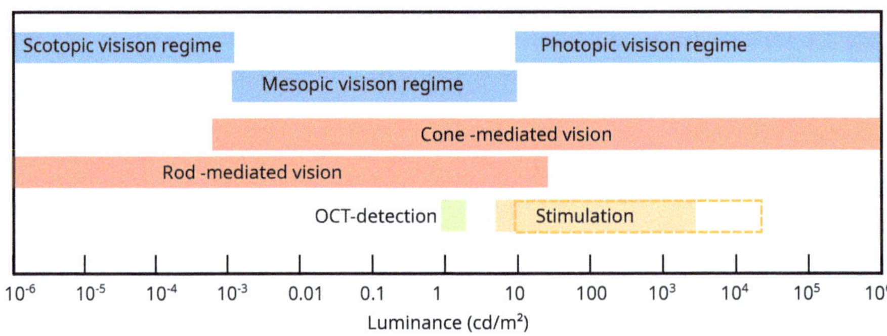

Figure 4.5: Regimes of scotopic, mesopic and photopic vision in luminance. Luminances for scotopic, mesopic, photopic, rod dominated and cone dominated vision were adapted by Zele and Cao [191]. The green bar corresponds to the luminance caused by the OCT-detection. The filled yellow bar corresponds to the calculated luminances for stimulation. The yellow dashed framed bar corresponds to the luminance considering the pupil diameter.

is used for the calculation with $V(830\,\mathrm{nm}) = 6.6^{-7}$ [99]. This results in a luminance of $2.2\,\frac{cd}{m^2}$. Therefore the stimulation of the OCT's light source is in the same range as the lowest used stimulation luminance ($7\,\mathrm{cd/m^2}$). A further reduction of the stimulation to investigate the functional respond in the scotopic regime is therefore not possible.

In order to measure the signal for varying stimulation intensities, for each luminance two different measurement were taken. Figure 4.6 shows the time courses corresponding to the average of these two measurements. From the different time courses it can be seen, that both, the optical path length change reached after 8 s as well as the initial slope vary with the stimulation luminances. Yet, for both parameters there seems to be no linear relationship to the luminances. Their dependency on the luminance is shown in Figure 4.7 a and b. The initial slope shown here was calculated by linear approximation of the optical path length change over the first 125 ms of stimulation. A Gaussian-curve was fitted to the data points for the maximum increase of the optical path length and an error-function for the initial slope and were added for better visualization of the data. In the range from 37 cd/m^2 to 1240 cd/m^2, there is an approximate linear relation between the logarithm of the stimulation luminance and, both, the initial slope and the functional change after 8 s. The functional signal increased by about 50 nm and the initial slope by 10 nm/s with a doubling of the luminance. However, for lower stimuli the two parameters nearly become independent on the luminance. In fact, there is hardly any difference in the time course between stimulation with 37 cd/m^2 and 7 cd/m^2. In addition, at very high stimulation levels (4600 cd/m^2), the initial slope saturates (Fig. 4.7 b) and the functional change after 8 s even decreases (Fig. 4.7 a).

The shape of the curves in Figure 4.7 a and b at lower stimulation levels can be understood by varying degrees of the contribution of rods and cones. The higher stimulation levels between 4600 cd/m^2 and 137 cd/m^2 are in the photopic range of vision. Here, mostly cones are involved in the generation of the visual signal. Thus, it can be assumed that the observed functional signal in that range is dominated by cones. The functional signal caused by the cone vision is reduced after 8 s stimulation by 50 nm when the luminance is halved. Assuming the contribution of the cone signal is further reduced by 50 nm, if the luminance is halved again, then a stimulus of 16.8 cd/m^2 would hardly leave any measurable contribution of the cone signal. Under this hypothesis, the signal detected from a stimulus of 16.8 cd/m^2 and lower would be dominated by rods.

The photocurrent in rods already reaches saturation at a photopigment bleaching of 0.05% – 0.1% (Sec. 2.6.1.2). This is reached for stimulation luminances close to 16.8 cd/m^2. The fact that no difference can be seen between the last three stimuli could therefore be explained by the fact that the signal is dominated by the functionality of the rods, which are still in saturation.

The saturation and the decrease of the curves in Figure 4.7 at higher stimulation can be related to light adaptation, since measurements were performed at a dark adaption state. Although, the human eye is able to detect stimuli over 12 orders of magnitude of luminance, it can only differentiate within a stimulation range of three orders of magnitude at any given time. The huge working range is achieved by adaption to dark or light conditions. The light adaptation process begins after about two seconds of stimulation and ensures that the photoreceptors become less sensitive to incoming light. One phenomenon observed during light adaptation is that the photoreceptor OS shortens over a period of several minutes [192]. The decrease in the time course of

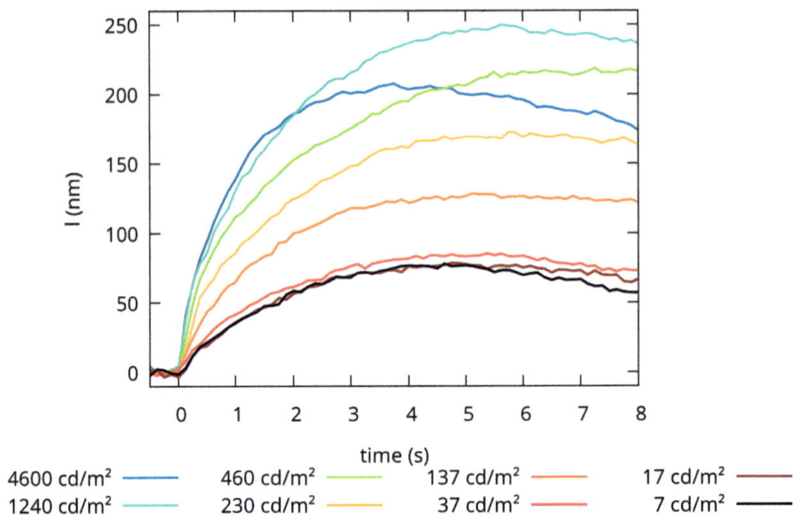

Figure 4.6: Time course of the OPL changes in the 3 mm periphery for a steady stimulation over 8 seconds with different intensities

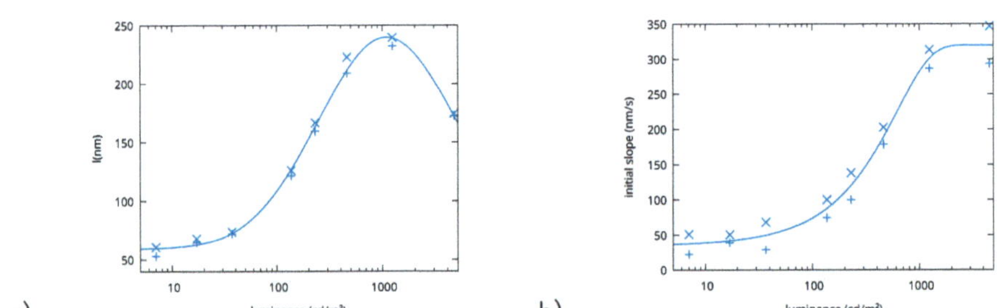

Figure 4.7: a) Elongation of the optical path length in the periphery after 8 seconds of continuous stimulation at different stimulation luminances. b) Initial slope of the optical path length.

the optical path length change at a stimulation of 4600 cd/m^2 after 2 s may therefore be caused by an incipient, opposite process of light adaptation, which is particularly pronounced as the stimulation intensity is outside the acute sensitivity of the eye.

4.1.1.5 Depth-Dependence of Functional Changes in the Photoreceptor OS

The evaluation of the optical path length between the IS/OS and the layer 5 pixels deeper (each averaged over 3 depth pixels) was chosen in such a way that a maximum change, as well as most stable and reproducible results were achieved. At this point, however, the differences of the optical path length change along the OS shall be investigated. In order to obtain a higher axial discrimination, the phase of each layer was averaged over only 2 pixels in depth and each layer was referenced to the layer just below it (see Fig. 4.8). This reduces the SNR of the functional signal slightly, but it is possible to examine three different layers in the OS separately. For the investigation two measurements from the last section (4.1.1.4) were selected; one stimulated with an intensity of 4600 cd/m^2 and one with an intensity of 1240 cd/m^2. Those two measurements were chosen in order to investigate the abnormal behavior in the time course of high stimulation luminance (meaning the saturation of the slope and the decrease of the optical path length after 2 s) in a depth dependent manner.

In Figure 4.9 (a-f) the en-face images of the functional change after 8 s stimulation for those two measurements are shown, in the three different depths. In addition, Figure 4.9 g shows the corresponding time courses of the functional signals. For a stimulation of 1240 cd/m^2, an increase in the optical path length is measured in all three layers, with the greatest change (approx. 70 nm) occurring in the central layer (green curve). This also applies to the functional change at a stimulation power of 4600 cd/m^2. While the curves in the lowest and uppermost layer are in the first two seconds comparable to those at lower stimulation, the central layer (dashed green line) shows an increased response in the first two seconds. After 2 s, the signal curve flattens in the central and lowest layer (green dashed and blue dashed curve) and falls below the corresponding curves at lower stimulation levels.

Particularly interesting are the matching en-face images. While a stimulus of 1240 cd/m^2 leads to a homogeneous increase of the optical path length over the stimulated area in all layers, this is only the case in the upper layer for a stimulus of 4600 cd/m^2. On the central and lowest layer (green and blue), on the other hand, an increase in the optical path length can only be observed in the marginal areas of the stimulation. In contrast, in the central area of the stimulation, there is hardly any change in the optical path length after 8 s, and in some places, even a reduction can be observed. The time course, however, shown in Figure 4.9 g correspond to a phase change averaged over both areas (marginal and central area), which is why the curves do not drop back to zero completely. It can therefore be assumed that the drop of the functional signal at a stimulation power of 4600 cd/m^2 is much more drastic than the

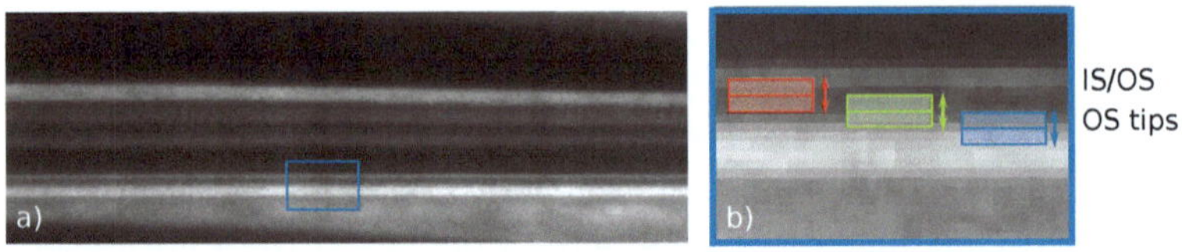

Figure 4.8: Phase evaluation for three different layers in the photoreceptor OS: a) B-scan of the volunteer retina in 3 mm periphery. b) Close up of the blue marked region in (a) the different colored layers indicate the three different layers for which an oven phase evaluation were done.

curve shown in Figure 4.6 suggests. The contraction observed here in the lower part of the OS for very high stimulation luminances can thus, probably, be assigned to the in Section 4.1.1.4 discussed shortening of the outer segment due to the light adaption process.

The inhomogeneous behavior of the functional changes over the stimulated area can be explained by a reduced stimulation in the marginal areas due to aberrations or movements during the measurements, so that the marginal regions do not exhibit the opposing process of contraction, caused by the light adaption process.

4.1.1.6 Reversibility

Finally, the reversibility of the functional signal was investigated using the same settings and stimulation parameters as in Section 4.1.1.4. However, at this point, the stimulus was not applied over the entire measurement period of 8 s but only for 100 ms. The corresponding optical path length changes over time are shown in Figure 4.10.

In the first phase of the response, there is a steep increase of the optical path length, followed by a slower decrease or recovery of the functional signal, finally dropping back to zero. While the initial slope decreases with decreasing stimulation luminances and becomes independent of the stimulation intensity for luminances of 137 cd/m^2 and below, the signal recovering of the different simulations has the same slope for all stimulation levels.

It can also be observed that the increase in the functional signals continues after the stimulus ends. The duration of the optical path length increase thereby varies with the stimulation level, too. Yet, the way the duration of the rising phase changes for different stimulation levels seems a little bit random at first. First, the duration of the signal's increase gets shorter with a decreasing luminance, but at a luminance of 460 cd/m^2, it suddenly becomes longer and then again decreases further.

However, the measured time courses can be described by a superposition of two independent curves, which means, that the behavior of the functional signal could be explained by a composition of two different physiological processes. A first or

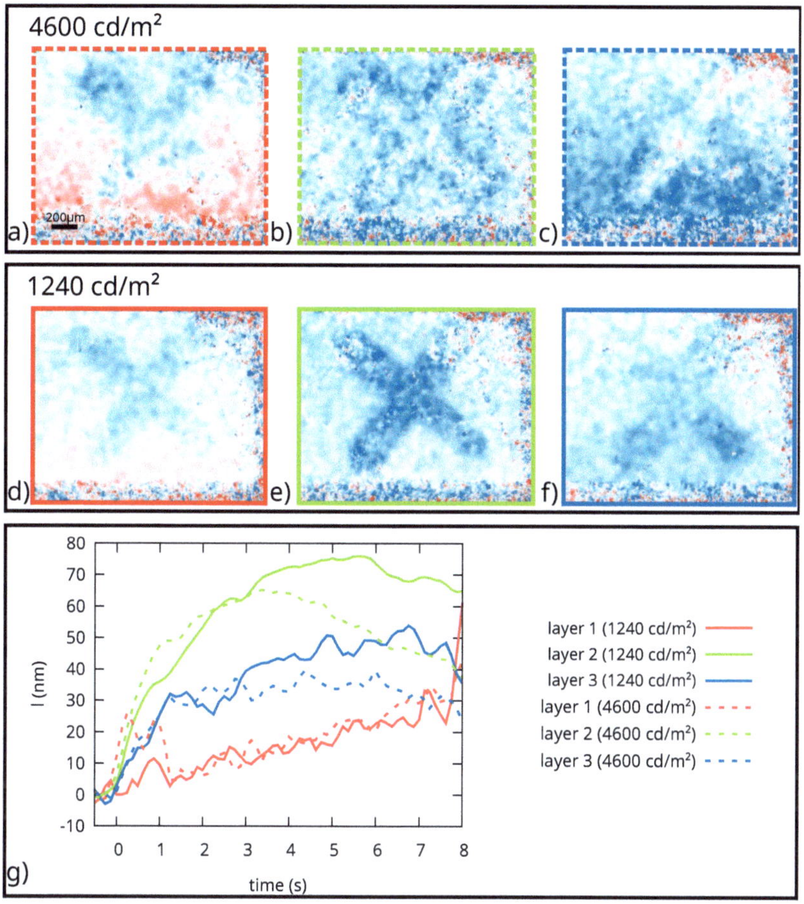

Figure 4.9: Evaluation of the optical path length change for different levels of the PR OS a)-c) En-face images after 8 s stimulation with a luminance of 4600 cd/m^2 in the first (a) second (b) and third (c) layer respectively. d)-f) En-face images after 8 s stimulation with a luminance of 1240 cd/m^2 in the first (d) second (e) and third (f) layer, respectively. g) Corresponding time courses of the optical path length changes.

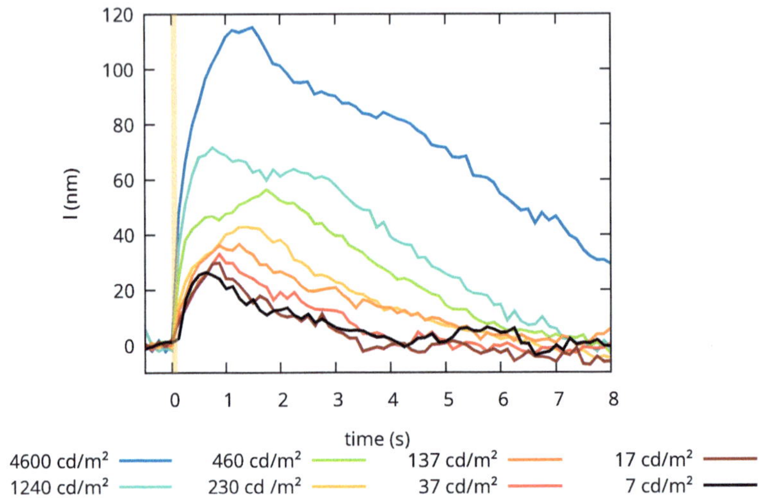

Figure 4.10: Time courses of the optical path length changes for stimuli of 100 ms duration and different luminances.

fast signal, which reacts very sensitively to the used stimulation level, and a second somewhat slower process, which reacts less sensitively to the used stimulation level. In order to better investigate the processes quantitatively, a sum of two impulse responses was fitted to the measured signals:

$$f(t) = a_1 t \cdot \exp\Big(-tb_1 - c_1 t^2 - d_1 t^3\Big) + a_2 t \cdot \exp(-xtb - d_2 t^3)$$

Here, the individual impulse responses should describe two different processes, where $a_{1,2}$ are parameters that model the respective initial slope, and $b_{1,2}$, c_1 and $d_{1,2}$ are parameters that model the respective duration of the rising phase and the decline of the curve. For the second, slower and longer lasting process, a quadratic term for the fit of an impulse response seemed unnecessary, which is why this term was omitted in the fit function. The resulting fit curves as well as the respective individual impulse responses are shown for the different stimulation luminances in Figure 4.11.

When the time needed to reach the maximum for the two processes is examined independently of each other, it can be seen that, for both processes, the duration of the signal rise is continuously shortened as the stimulation luminance decreases. However, for the second, slower process this change in duration is significantly stronger (see Fig. 4.12 a).

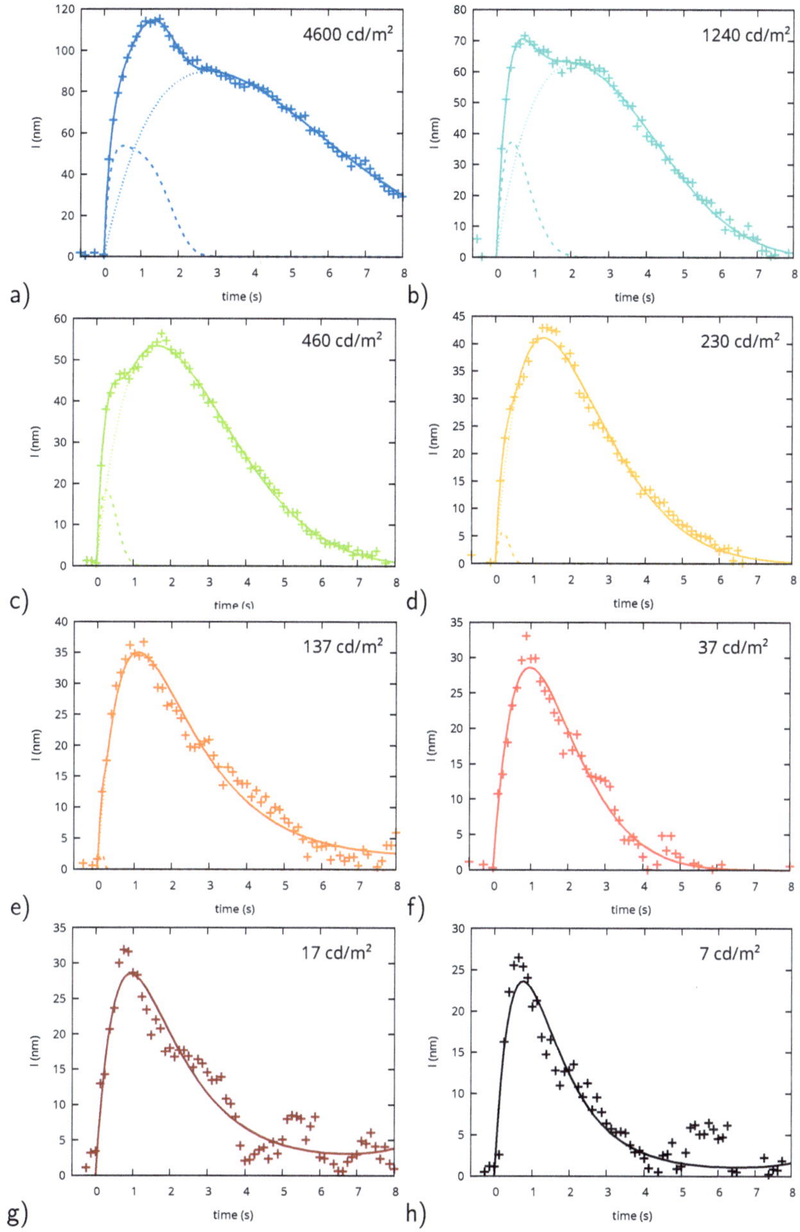

Figure 4.11: Fit of the functional signal to a short stimulus (100 ms) with different intensities. Measured change in the optical path length (crosses) , the fit curve, which describes the signal as the sum of two impulse responses (continuous line), the first impulse response (dashed lines), and the second impulse response (doted line) are shown.

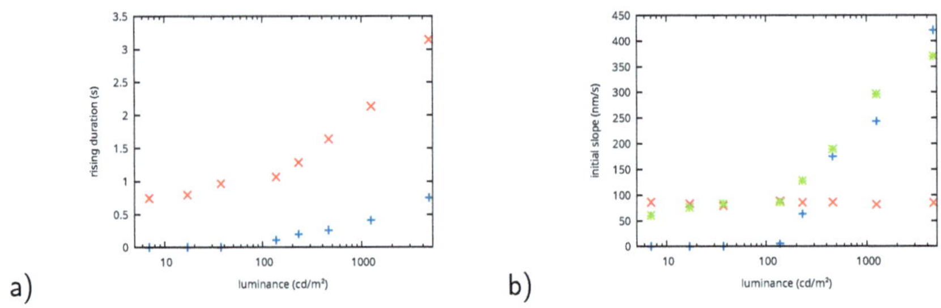

Figure 4.12: Dependencies on the stimulation intensities. a) Duration until the maximum is reached in dependency of the used luminance for the two different impulse responses of the fit curve, first impulse response (blue), second impulse response (red). b) Initial slope of the two impulse responses in dependency of the used luminance, first impulse response (blue), second impulse response (red), initial slope obtained by the measurements itself by linear extrapolation (green).

In the initial slope, however, the two processes behave differently. While the slope of the first process also decreases as the stimulation luminance decreases, the slope of the second process is more or less constant for all investigated luminances (see Fig. 4.12 b).

The green values in Figure 4.12 b correspond to the measured slopes obtained by a linear approximation of the optical path length change within the first 125 ms after stimulation. This means that the slope of functional signal is dominated by the first process at high luminance (4600 cd/m^2 and 1240 cd/m^2), but for weaker stimuli, this process quickly becomes smaller so that the second process dominates the functional signal. For the weakest three luminances (37 cd/m^2 to 7 cd/m^2), only the second process is visible. In this case, only one impulse response was fitted to the measured data. These single-response luminance levels are at the same luminance for which for continuous stimulation (Sec. 4.1.1.4) a stagnation of the slope and the signal intensity after 8 s was observed. In Section 4.1.1.4 it was argued that this may be caused by a dominance of the rod signal over the cone signal. This explanation also fits well with the data presented here.

In this case, the first, fast process would be caused by the cones, which can still perceive the visual stimulus with sufficient contrast at the used luminance and therefore can still vary their slope accordingly to the stimulation intensity. The rods, on the other hand, are in saturation of their photocurrent and therefore cannot respond to the stimulus intensity with a varying slope. Also, the duration of the signal's increase and the total duration of the signal response is much shorter in the first process than in the second process, which also fits well to the hypothesis that the two processes are caused by cone and rod signals, as cones react much faster to changes in the signals than rods.

4.2 Intrinsic Optical Signals in the Inner Plexiform Layer

Due to the volumetric recording, it is possible to perform a functional evaluation of the photoreceptor layer as well as other neuronal layers during a single measurement. Besides the photoreceptor OS, the IPL is the most suitable layer for that purpose, since it provides the highest signal intensity in the OCT images, other than the RPE, the IS/OS and the NFL. In addition, it includes the functional connection of the bipolar, amacrine, and ganglion cells, which makes it particularly interesting from a physiological point of view. Therefore, in this section, the functional evaluation in the IPL will be examined more closely. For the functional phase evaluation of this layer, the upper edge of the IPL is segmented, and different layers are determined for the phase referencing that are parallel to this segmented upper edge. The choice of the thickness and distances of those layers was varied depending on the aspect of the signal to be investigated.

4.2.1 Optimal Reference Layer

When using the IPL for functional phase evaluation, the optimal reference layer is unclear. Since the IPL includes the synaptic contacts of amacrine, bipolar, and ganglion cells, a variety of of processes could cause a functional signal, and accordingly, the optimal reference layer may differ. For example, if a functional change is caused by an increased metabolism in the activated cell, then a reference to the corresponding cell body would be optimal since most important metabolic processes take place there. Accordingly, if a functional signal is caused by an increased metabolism of the bipolar or amacrine cells then referencing to the INL would be most useful, since this is where their cell bodies are located. If the signal is caused by an increased metabolism of the ganglion cells, a reference to the GCL would be most suited. In addition, it may also be that the signal is caused by the signal transduction in the IPL, for example, by ion flux or potential change in the post-synaptic membrane or by the release of neurotransmitters from the pre-synaptic membrane. In this case, it would be best if the two reference layers were both selected from the IPL.

For this reason, in the first step, the optimal reference layer was determined. Different reference layers and their influence on a functional change were investigated. For this investigation, a retinal area in 0.8 mm temporal of the fovea was stimulated, continuously over the whole measurement duration (8 s) with an x-shaped pattern with a luminance 137 cd/m^2.

To investigate the signal with the different potential reference combinations, the phase in the IPL was averaged over 7 pixels and then referenced to the ganglion cell layer (axial averaging of the phase over 5 depth pixels) (Fig. 4.13 b marked in yellow)

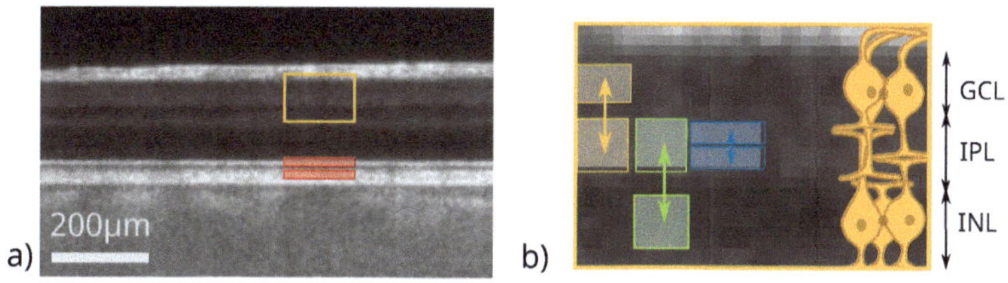

Figure 4.13: Different reference layers for the functional phase evaluation. a) B-scan of the volunteer's retina in 3 mm periphery. The red marked region corresponds to the reference layers for the photoreceptor cells. b) Close up of the yellow marked region in (a); the different combinations of reference layers are marked in different colors: yellow GCL-IPL, (green) IPL-INL and (blue) IPL-IPL.

or to the bipolar cell layer (axial averaging of the phase over 7 pixels) (Fig. 4.13 b marked in green). Furthermore, the upper half of the IPL (averaged over 3 pixels in depth) was referenced to the lower half (averaged over 3 pixels in depth) in order to be able to estimate functional changes that are only caused by the synaptic connections in the IPL (see Fig. 4.13 b marked in blue).

A corresponding en-face image of the optical path length change for the different references after 8 s is shown in Figure 4.14. The functional change in the photoreceptor OS is shown for comparison in Figure 4.14 a. When referencing the IPL to the bipolar cell layer there is no area in the field of view, which behaves particularly different than the rest of the field of view (Fig. 4.14 b), which means there is no functional change in the phase observable. However, when referencing to the ganglion cell layer a deformed x-shaped pattern appears with an increased optical path length (Fig. 4.14 d). Similarly, an increase of the optical path length can be seen at the same location, when the upper half of the IPL is referenced against the lower half (Fig. 4.14 c). In order to study the functional response quantitatively the phase was averaged over the area in which the functional signal were observed. However, unlike in the photoreceptor OS, in the IPL additional movements are present, for example through vascular pulsation and other metabolic processes. Thus, a continuously changing and non-reproducible background is superimposed on the functional signal. In order to reproducibly detect the time course of the functional signal, the phase change of the background was also measured and subtracted from the functional signal. For this purpose, the phase was averaged in a second area as far as possible from the functional signal. The two areas, in which the functional signal and the background were averaged are shown in Figure 4.15.

The corresponding time courses are shown in Figure 4.16. Despite subtraction of the background, the two time courses are overlaid with pulsation artifacts. These cannot be removed completely, because the pulsation affects distinct areas of the imaged field

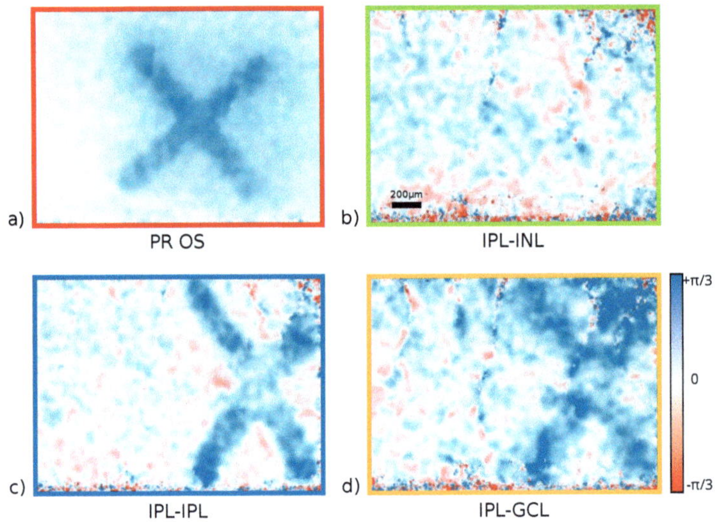

Figure 4.14: En-face images of the functional changes in the photoreceptor OS an the upper retina for different reference combinations. a) Functional change in the PR OS, b) functional change between the IPL and the INL, c) functional change between the upper half of the IPL and the lower half of the IPL, and d) functional change between the GCL and the IPL.

of view differently. The pulsation artifacts in the phase evaluation of the GCL-IPL referencing are larger than the artifacts of the phase evaluation between the different layers of the IPL, since the larger retinal vessels are located in the GCL, while in the IPL smaller capillary vessels are present, which only cause smaller artifacts. Apart from the pulsation artifacts, the basic increase in optical path length over time is approximately the same in the evaluation of the GCL-IPL and IPL-IPL referencing. This suggests that the ganglion cell bodies hardly contribute to the signal, since including them does not affect the measured phase changes. Another indication that the functional change has its origin in the IPL is the en-face images (Fig. 4.14 c-d). While the lateral appearance of the functional changes is sharply limited in the phase evaluation of IPL-IPL, the appearance in the phase evaluation between GCL-IPL is much more blurred, which could suggest that the signal observed between GCL-IPL are only a shadow effect from the changes between the IPL-IPL.

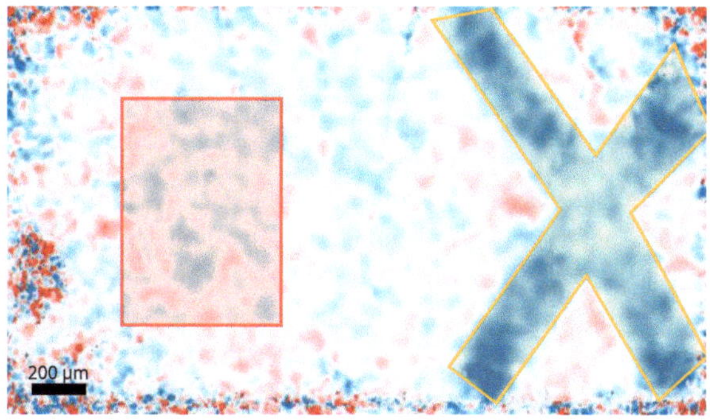

Figure 4.15: Exemplary areas used for background correction. The yellow marked area is the mask for the functional changes that is used to extract time courses; the red marked area corresponds to the mask used to obtain background noise.

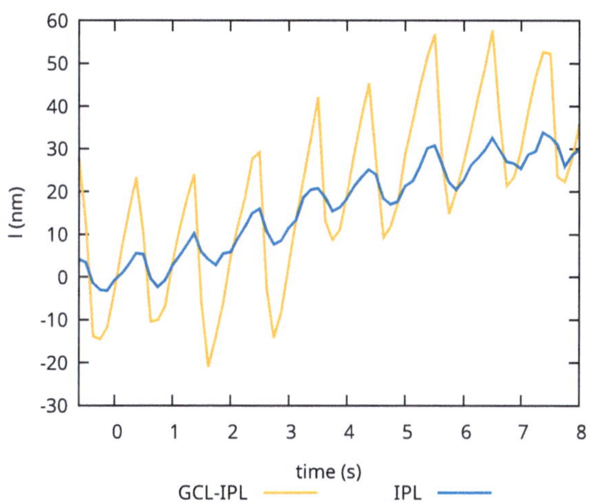

Figure 4.16: Time course of the functional change in the upper retinal layer. Functional change between the GCL and the IPL (yellow) and functional change between the upper half of the IPL and the lower half of the IPL (blue).

4.2.2 Investigation of the Signal Dependency of Different IPL Levels

According to the previous section, the majority of the functional signals originate from the IPL. In the following, the depth dependency of the functional signal within the IPL will be investigated in more detail in order to investigate if the functional changes are particularly pronounced in a certain level of the IPL or if certain levels are particularly sensitive to a specific feature in the visual signal. For this purpose, measurements were made close to the fovea (0.8 mm temporal). The depth resolution of the FF-SS-OCT only allows differentiation between 3 different layers in the IPL. For each layer, an average of 2 pixels in depth was referenced to the layer directly below this layer (also averaged over 2 pixels), as shown in Figure 4.17 b. For the evaluation, two different stimuli were used: first, a continuous stimulation over 8 s with a luminance of 137 cd/m^2 and, second, a 10 Hz flicker stimulation (50% duty cycle) with a peak luminance of 137 cd/m^2.

The en-face images of the corresponding phase changes after 8 s are shown in Figure 4.17 (c-h) and the respective time courses of the phase changes of these three layers are displayed in Figure 4.17 i. First of all, it can be seen from the continuous stimulation, that the different levels of the IPL respond fundamentally differently to the same visual stimulation (Fig. 4.17 i continuous lines). While in the top layer (closest to the ganglion cell layer) there is a continuous increase of the optical path length change, which reaches approximately 30 nm after 8 s, the seconds or central layer (green) hardly shows any change in the signal, and in the bottom layer (blue), which is closest to the bipolar cells, a continuous decrease over 8 s can be observed, reaching an optical path length change of approximately −20 nm.

However, when the retina was stimulated with a 10 Hz flicker stimulus over 8 s with the same intensity that was used for the continuous stimulation, a different signal response can be observed. The respective time courses are marked by the dotted lines in Figure 4.17 i, and the corresponding en-face images after 8 s are displayed in Figure 4.17 f-h. While the signals in the uppermost layer (red dashed) and in the lowest layer (blue dashed) hardly change compared to the respective signal for the continuous stimulation, a clear signal can now also be seen in the central layer (green dashed) in the en-face image. In the time course, it is also clearly visible that there is a continuous expansion of the central layer, which reaches a value of 30 nm at the end of the measurement and is therefore about the same size as the signal from the upper layer.

The different response at different depths of the IPL can be explained by the interconnection behavior of different bipolar and ganglion cell types. As already mentioned in Section 2.6.1.4, there are different types of bipolar and ganglion cells, each of which is sensitive to certain characteristics of the visual stimulus. A special characteristic of the different ganglion and bipolar cell types is, that they only form their synaptic connections at certain depth of the IPL (see Fig. 4.17 b), which means that a certain visual feature is only processed at a certain depth of the IPL. Basically, a distinction can be made between 5 different layers of approximately the same thickness in the IPL [126]. Due to this interconnection characteristic, it can be assumed that contributions of distinct cell types can be differentiated by evaluating the phase changes for different depths of the IPL independently. Although the axial resolution used here did not resolve 5 but only 3 different layers in the IPL, the depth-dependent functional responses to different visual stimuli fit surprisingly well with the current knowledge of how different visual features are processed in the IPL.

The upper two layers (red and green), which are investigated here correspond to the region in the IPL in which ON bipolar cells connect to ON ganglion cells, whereas the lower layer (blue) is the region in which the OFF bipolar cells connect to OFF ganglion cells [193]. In principle, OFF cells react to visual stimulation in the opposite way to ON cells, i.e., visual signals, which stimulate certain ON cells, inhibit the corresponding OFF cells, and vice versa. The phase signals shown here are in general consistent with this interconnection behavior. The upper two layers show an increase of the optical path length during light stimulation, while the lower layer (blue) shows an inverse reaction, i.e., a shortening of the optical path length.

Furthermore, it is known that cell types possessing synaptic connections with ionotropic receptors, i.e., receptors that react quickly to changes in the signal over time, are preferentially formed in the center of the IPL, while synaptic connections with metabotropic receptors, meaning receptors which only react slowly to temporal change, are more likely to be found at the outer edges, i.e., near the adjacent cell layers, which is why cells that form their synaptic connections in the middle of the IPL are sensitive to temporal changes in the visual signal [119].

This wiring scheme also fits well with the measurements shown here, which shows an increased sensitivity of the phase in the central part of the IPL (green curves) to temporal changes in the stimulus, while the outer two layers (red and blue) do not seem to be affected by temporal changes in the visual stimulus.

The measurements shown here therefore suggest that differentiation of the contributions of different functional cell types by depth-dependent investigation of the phase signals is possible, without resolving single ganglion or bipolar cells.

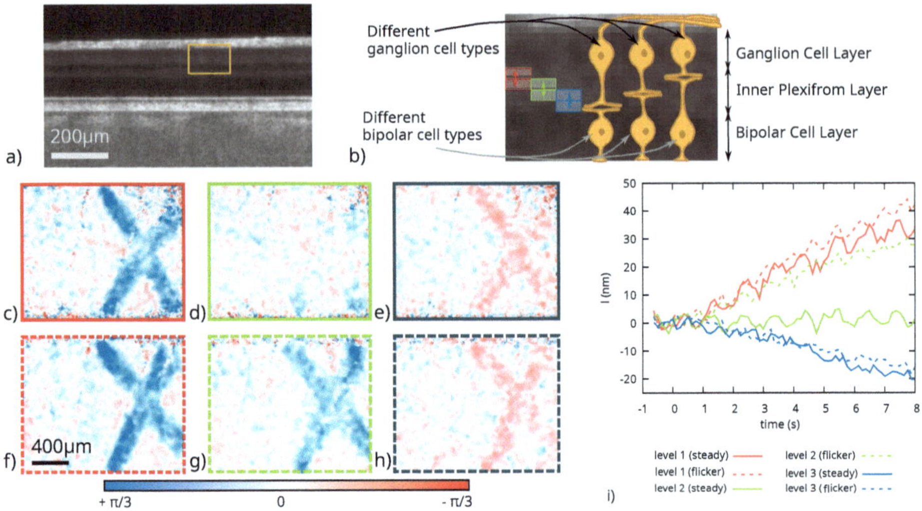

Figure 4.17: Investigation of the depth dependency of the functional phase changes in the IPL. a) B-scan of the volunteer. b) Closeup of the yellow marked region in (a) that illustrates the connection behavior of the depth dependent stratification of different bipolar and ganglion cell types. The three different layers of the IPL, which are evaluated separately, are marked in different colors. c-e) En-face images of the phase change after 8 s continuous stimulation for the different IPL layers. Frame colors correspond to the regions marked in (b). f-h) En-face images of the phase change after 8 s flicker stimulation with 10 Hz for the different IPL layers. Frame colors correspond to the regions marked in (b). i) Corresponding time course of the functional phase changes.

4.2.3 Investigation of the Center-Surround Signal Transduction of the Receptive Field

In addition to the axial dependency, the lateral dependency of the signal will now be investigated. For such an analysis the signals from the uppermost layer in the IPL, closest to the ganglion cells, will be used, since the functional changes can be observed most clearly in this layer. From such an en-face image it can be observed that the functional changes are differently pronounced for different regions in the 'x'. More precisely, a higher signal can be observed in the arms of the 'x' (Fig. 4.18 marked with green arrows), than in the central area (Fig. 4.18 marked with yellow arrow). Such an inhomogeneous behavior over the stimulated area could be caused by an increased sensitivity for edges. Yet, the x-shaped stimulus has a very high edge to area ratio, so

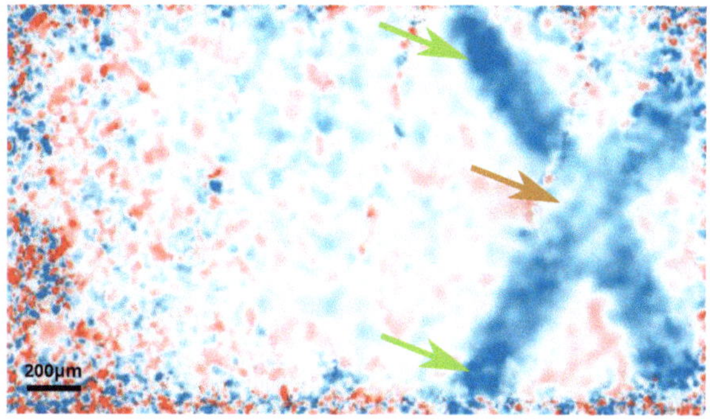

Figure 4.18: Inhomogeneous response within the activated area: En-face image of the functional change in the upper IPL after 8 s continuous stimulation. The functional response in the arms of the x-pattern (green arrows) is noticeably stronger than in the center of the activated area (yellow arrow).

depending on the size of the width of the edge filter it is difficult to distinguish between the functional signal generated at an edge and the functional signal generated by an even stimulation.

In order to verify, that there is an increased functional change at the edges of the visual stimuli the x-pattern was replaced by a square pattern, as this has a much lower edge to area ratio. Again, a standard stimulation (8 s, 137 cd/m^2, 3 mm periphery) was used. The corresponding en-face images for the functional signal in the photoreceptor OS and the IPL are shown in Figure 4.19. In addition to the en-face images, the profile of the phase change corresponding to that en-face images are shown in Figure 4.20.

Here, the phase information was averaged over the marked area in the y-dimension. It can be seen that the photoreceptors respond homogeneously over the entire area of the stimulation-pattern. The slight increase of the optical path length in the direct vicinity of the stimulated pattern is probably caused by aberrations of the projection of the stimulation pattern on the retina.

However, in the IPL the changes in phase are very inhomogeneous over the stimulated pattern. There is a particularly strong phase change at the edges of the stimulated pattern (Fig. 4.20 b marked with the blue arrows), while the phase change within the stimulation pattern is significantly weaker (Fig. 4.20 b marked with the green arrow). However, the phase change in the center of the stimulation pattern does not drop completely to zero but stays slightly above the baseline of the non-activated areas. Furthermore, it seems that in the direct neighborhood of the stimulated area in the IPL, the optical path length decreases over time, meaning the phase change becomes negative (Fig. 4.20 b marked with the red arrow). By this inhomogeneous behavior

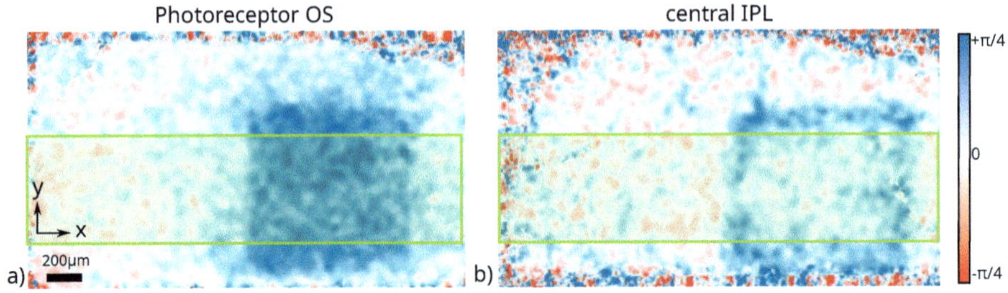

Figure 4.19: En-face images of the phase change caused by a square pattern stimulation. a) Phase change of the photoreceptor OS after 0.5 s of stimulation. b) Phase change of the IPL of the same measurement after 8 s of stimulation.

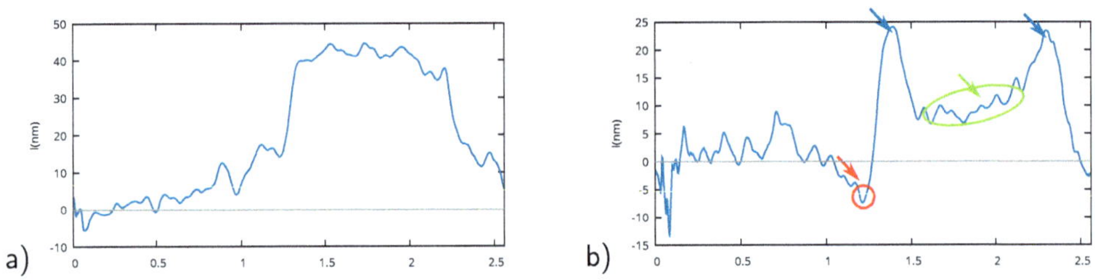

Figure 4.20: Profile of the phase changes along the x-axis of the imaged field of view. The profile were obtained from the en-face images shown in Figure 4.19 the phase were averaged in y-dimension over the colored region. a) Profile of the phase changes in the photoreceptor OS. b) Profile of the phase change in the IPL.

of the phase change, the contrast at the edge regions of the stimulation pattern is particularly increased, the contours become more prominent than in the photoreceptor OS, although the signal intensity is higher in the photoreceptor OS.

This inhomogeneous behavior of the functional change across the stimulated area can be explained with the center-surround processing within the receptive fields of some bipolar and ganglion cell types, as it was described in Section 2.6.1.6. In this processing scheme, stimulation of the marginal (surround) region of the receptive field acts antagonistically to stimulation in the central region.

Areas in which the center of the receptive field and only parts of the antagonistic surround are stimulated with light (edges) produce a greater functional change than areas in which both the entire center and the entire surround are stimulated. The fact that a functional change can be detected within the stimulation pattern, although significantly weaker than at the edges, is therefore either because the center activation has a higher influence than the surround inhibition or that additional cells that do not have such center-surround processing within their receptive field, contribute to the functional signal observed here. Finally, the decrease in the optical path length around the stimulated area could be caused by cells, which were only stimulated with light in their surround areas, and therefore are only experiencing an inhibition by the stimulation.

4.2.4 Frequency Dependence

In Section 4.2.2 it was shown that the phase changes in the central part of the IPL are sensitive to temporal changes in the visual signal. In this section, the frequency dependence of the signal changes in the central level of the IPL were investigated. In the last Section it is shown that there are also lateral differences in signal intensity (i.e., increased phase changes at the edges of the stimulation pattern). To ensure that these different signal contributions are not mixed, a distinction is made between phase changes at the edges of the stimulation pattern and phase changes in the central area of the stimulation pattern. For an optimal discrimination between these contributions, the square stimulation pattern introduced in the last Section was used. The investigation showed that the frequency dependence of the signal was only reproducible in the central area of the stimulation pattern. The edge region, on the other hand, reacts very sensitively to slight variations of the mask over which the phases were averaged. For this reason, the contribution of the edges was not further investigated.

The stimulation was done with a peak luminance of 137 cd/m^2 at 11 different temporal frequencies ranging from1 Hz, to 55 Hz and for a continuous illumination. All stimulation frequencies had a duty cycle of 50% and, thus, the number of photons stimulating the retina stayed the same for the different frequencies of the flickering illumination.

To evaluate the sensitivity of the phase change in the IPL with respect to the temporal change in the stimulation, the phase change reached after 8 s is investigated. To reduce the influence of noise or artifacts caused by the arterial pulsation, the resulting time course of the signal was smoothed with a moving average over 5 data points (corresponding to 625 ms). Figure 4.21 shows the response of the central part of the IPL for the different stimulation frequencies. For each frequency, two individual measurements were made, where each point in Figure 4.21 corresponds to one measurement. It can be seen that for a continuous stimulation or low stimulation frequency of 1 Hz there is only a minimal change in the optical path length of about 10 nm. However, the signal increases for higher stimulation frequencies (2 Hz – 6.7 Hz) and reaches a maximum change in the optical path length around 20 nm for stimulation frequencies of 6.7 Hz – 25 Hz. For higher frequencies, the sensitivity decreases again and reaches an optical path length change comparable to a continuous stimulation at a stimulation

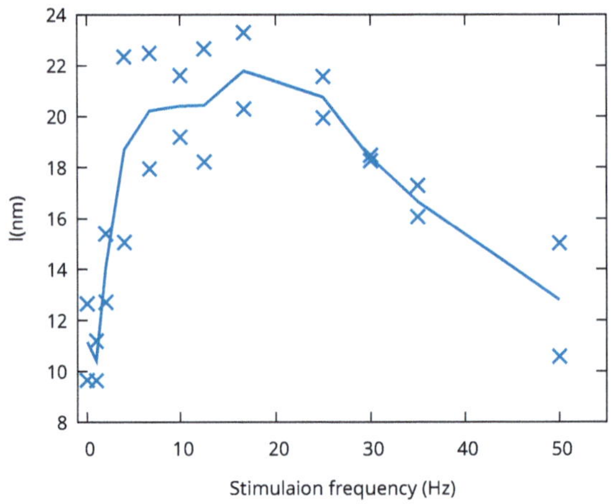

Figure 4.21: Frequency dependence of the functional phase change in the IPL. The optical path length changes reached after 8 s were evaluated. Each point correspond to an individual measurement. The continuous line connects the average of the measurements.

frequency of 50 Hz. The temporal dependence of the IPL's phase change observed here corresponds well to measurements using which measured electrical changes for different flicker stimulation in macaque retina [194, 195], and studies, which measured changes in the blood flow for different stimulation frequencies [29]. All observe a decrease in the functional contrast for temporal frequencies about 30 Hz and above .

4.2.5 Functional Connection of the Photoreceptor Layer and the IPL

A great advantage of the tomographic character of OCT is that functional changes over the full retina can be detected simultaneously. From Figure 4.14 it can be observed that the same visual stimulation leads to functional changes in the photoreceptor cell layer and the IPL, that are laterally shifted to each other. The fact that the signals in the photoreceptor layer and in the IPL appear at laterally different locations within a single measurement not only proves that these two signals have a different physiological origin but also provides the opportunity to investigate the functional connection of these two layers in detail. We already published the main results of this section [196].

In order to analyze the functional connections depending on the position of the retina, the changes of the optical path length were investigated for different positions in the retina. Starting directly in the fovea (0.8 mm temporal) and ending in the periphery at 3.3 mm superior the fovea. The stimulation was continuously over the whole measurement duration of 8 s using an luminance of 4600 cd/m^2. The phase was evaluated in this case between the upper and lower border of the IPL as described in Section 4.2.1. All data shown in this section were, in contrast to the results in all other sections, obtained with the Mach-Zehnder based FF-SS-OCT setup (Sec. 3.1).

Phase differences for three positions on the retina are shown in the three different rows of Figure 4.22. In order to analyze the functional connection between the photoreceptor and the upper neuronal cells, the x-pattern used for stimulation was manually outlined in the en-face images of the optical path length change for both the photoreceptor layer and the IPL. For this purpose, 12 points at the corners were set for each x (shown in the first two columns of Fig. 4.22). To determine the lateral offset of the two signals, the corresponding corner points of the two borders were connected (Fig. 4.22 third column).

The functional connection of the two layers can then be analyzed according to the length and the direction of these resulting connecting lines. The obtained offsets are displayed according to their retinal position in the SLO image in right column of Figure 4.22.

The measurements show a characteristic circular offset around the fovea. As expected from the anatomy of the retina, the signal in the IPL is further shifted outwards from the fovea than the signal in the photoreceptor layer. It can also be seen that the offsets between the two neuronal layers are greatest near the fovea (650 µm at 0.8 mm temporal) and decrease with increasing distance to the fovea. The smallest offset measured here is 180 µm at 3.0 mm in the periphery. Due to the radial outward shift, the signals in the IPL are deformed with respect to the stimulation pattern. This deformation is also strongest near the fovea due to the connection characteristics. The nature of the deformation also results in the functionally active area in the IPL being much larger near the fovea than in the periphery, although the area of stimulation and photoreceptor response are the same.

The observed interconnection behavior reflects the structure of the retina. To optimize the optical imaging in the fovea, the higher neuronal layers, such as the INL and the GCL, are missing. The ganglion and the bipolar cells, which are connected to the photoreceptors in the fovea, are accordingly shifted to the side, which is why such a large offset prevails here becoming smaller with increasing distance from the fovea.

The observed distances for the different positions in the retina of the photoreceptor and connected bipolar and ganglion cells fit well with histological examinations [197]. Besides the qualitative interconnection behavior, the magnitudes of the offsets agree surprisingly well with the ones previously measured and published (406 – 632 µm at 0.85 – 1.348 mm temporal [197]).

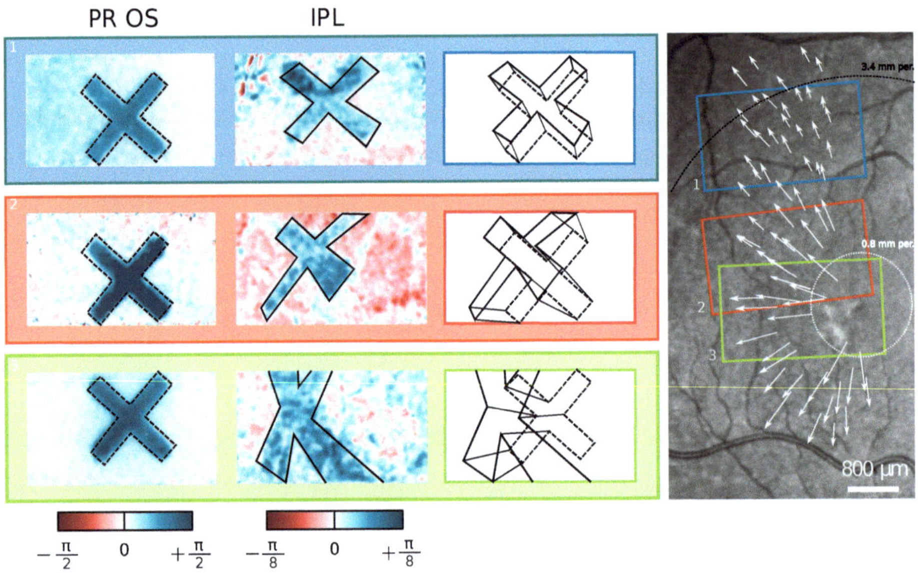

Figure 4.22: Map of the functional connection between photoreceptor layer and IPL. The three different colored rows show the functional signal for different positions in the retina. The corresponding positions are marked with frames in their respective color in the SLO image (right). First image in each row display an en-face image of the functional change in the PR OS, second image is a en-face image of the functional change in the IPL, and the third image corresponds to the functional connection of these two signals.

5 Discussion

In the last chapter, it was shown that the phases of the OCT signal can be used to visualize functional changes in the retina. Furthermore, the physiological significance of the observed phase changes in the OCT signal was investigated. In order to assess the importance of this new imaging contrast, two different aspects will be discussed below. First, the functional phase imaging using FF-SS-OCT will be compared to other contrast methods for functional retinal imaging, especially to other phase-sensitive OCT systems, and secondly, the possible molecular origins of the phase change will be discussed.

5.1 Comparison of Contrast Mechanisms for Functional Retina Imaging

A number of different imaging modalities have addressed functional contrast in the retina in the last years. So far, however, the only method for measuring retinal function that has become established in clinical practice is electroretinography (ERG). The ERG measures the electrical excitation of the entire retina and thus, detects an integrated signal on the activity of all activated retinal neurons, which allows to a limited extent discrimination of the contributions of different cell classes, but not the activity of individual cell types or cells.

Similarly, functional imaging via the blood flow was demonstrated in humans [20] and animals [28, 29]. Although, detection of perfusion changes is easily realized and does not burden the patient, the change in blood flow again only reflects an integrated signal about the activity of the entire retina and is not able to differentiate the contributions of different cell classes, cell types, or individual cells.

Rhodopsin densitometry [34, 35, 37, 38, 198] is able to clearly assign the changes measured to the photoreceptors layer and is even able to detect the functional changes for individual cells [37, 38], but the functional contrast is limited to the photoreceptors and cannot be used to visualize activity in other cell classes such as bipolar cells or ganglion cells.

Autofluorescence imaging can in general detect metabolic function in cells. E. g., using two-photon excitation, formation and clearance of retinol was measured in individual cones [43]. Although the functional contrast of neuronal cells via fluorescence imaging has been demonstrated ex-vivo and in animal studies, so far it has not been shown for in-vivo studies in humans. Additionally, it is not a tomographic imaging method, which means that different layers of the retina cannot be imaged for functional activity at a time.

Finally, there is functional retinal imaging based on intensity fluctuations of scattered light measured with SLO and OCT. Here, functional changes could also be assigned to individual levels in the retina and also to individual photoreceptor cells [45,51,52,55]. Furthermore, OCT measurements are tomographic, which means that they are able to detect functional changes in several levels of the retina simultaneously [56,57]. However, the functional contrast suffered from poor SNR and the amplitude fluctuations were not sensitive enough to detect functional changes in the retina quantitatively combined with a good lateral resolution. Therefore, the functional contrast based on intensity changes did not provide reproducible time course, or the time course was only extracted statistically over a large area of the retina.

Phase-sensitive OCT, shown in this thesis, is able to resolve the functional changes in all three spatial directions. Thus, not only the contributions of single cells within one depth level can be measured, e.g., the contributions of single PR, but also different retinal layers can be investigated simultaneously. This allows the simultaneous measurement of different cell classes and also enables the analysis of the functional connection of those layers. In addition, due to the high sensitivity of the phase to tissue changes, very small changes in the retina of a few nanometers can be detected, whereby a good SNR of the functional changes is achieved. Thus, quantitative measurements of transient changes are possible.

5.1.1 Comparison of Different Phase-Sensitive OCT Techniques

Functional phase contrast for photoreceptor cells was first demonstrated by our group in 2016 [190]. However, the proof-of-principle results of this first study are not shown here, because since then the post-processing has fundamentally evolved. The results presented in this thesis essentially reproduced our initial findings and extended them beyond the first proof-of-principle and provided insight into the physiological significance of the measured signals. In the meantime, phase-sensitive OCT for functional imaging of retinal neurons was also demonstrated by line-field OCT and different scanning OCT systems.

Following, the results shown in this work will be compared to the results generated by our group in 2016 and published results, which were obtained with other OCT systems, starting with a short overview of the different OCT systems.

FF-SS-OCT The OCT setup used to generate the results in 2016 is nearly exactly the Mach-Zehnder setup described in Section 2.4.3. Only the stimulation unit for the eye was changed. Instead of using an LED combined with a pattern mask, the retina was stimulated with a modified LED LCD projector [190]. This provided much higher stimulation intensities of up to 19 µW/mm^2, which is about an order of magnitude more than the stimulation intensity of the experiments in this thesis. Thus, stimulation was far in the upper photopic vision. The specification of the FF-SS-OCT setup used in 2016, are listed in Table 5.1.

Additionally, the post-processing of the recorded data was less sophisticated. The registration considered only axial and lateral translations of the data. Therefore, the movements within a measurement could be corrected with sufficient stability for only a few seconds. Over longer measurement times, slight rotations and deformations of the retina occurred, so that only correcting translation was not sufficient to register the speckle patterns over the entire image field. For this reason, functional examinations of the retina were only possible up to 3 s. Since the maximum number of volumes that can be recorded by FF-SS-OCT is limited by the built-in memory of the high-speed camera, a reduction of the total measurement time is accompanied by an increase in the volume rate. Two different volume rates were used for the functional evaluation in the previous studies. One is a volume rate of 20 Hz (total measurement time of 3 s) and the other is a maximum volume rate of 167 Hz (total measurement time 420 ms). The volume rate of 167 Hz corresponds to the maximum volume rate that can be achieved at the given depth resolution and is limited by the sweep rate of the used broad-sweeper.

For calculating the phase changes, a simple temporal referencing of the phase to pre-stimulus volumes was used. Since the influence of stochastic changes was not minimized compared to the systematic changes, the functional phase evaluation was stable only for a very short period of time, especially in areas where a lot of additional movement took place, e.g., in areas close to vascular networks, before the speckle patterns were decorrelating and a meaningful phase evaluation was no longer possible. This basically prevented a useful evaluation of the IPL at that time.

Scanning OCT Shortly after the first demonstration, the functional phase-sensitive contrast in the retina was also shown with scanning OCT systems. Scanning systems, unlike full-field systems, do not require a high-speed area camera, but only a photodiode or a line detector. Therefore, they have the advantage of a real-time data readout and the data acquisition time is not limited by the memory chip of the camera. However, they have to provide fast and stable scanning in both x- and y-direction to prevent that the recorded volumes are corrupted by gaps or motion artifacts of the eye.

So far, the phase-sensitive functional contrast was demonstrated by two different research groups using scanning OCT.

	FF-SS-OCT (2016)	FF-SS-OCT (now)	SS-OCT [65,199]	SD-OCT [200]	LF-SD-OCT [201]
max. volume rate	167 Hz	167 Hz	32 Hz	10 Hz	324
max. B-scan rate	—	—	–	3 kHz	16 kHz
max. measuring duration	3 sec	8.625 sec	15 sec	5 sec	1.2 sec
need for area camera	yes	yes	no	no	yes
need for scanning devices	no	no	yes	yes	yes
λ_0	840 nm	840 nm	1060 nm	790 nm	840 nm
$\Delta\lambda$	50 nm	50 nm	80 nm	42 nm	50 nm
axial resolution	6.2 µm	6.2 µm	7.8µm	4.7µm	4.5µm
additional setups	no	no	AO+SLO	AO	AO+LSO

Table 5.1: Specifications of different phase-sensitive OCT systems. The values of the maximum volume rate, maximum B-scan rate and maximum measuring duration correspond to the respectively highest value shown by the research groups for a functional evaluation in the photoreceptors and not necessarily correspond to a limitation of their systems. For the SD-OCT system and the line-field (LF) SD-OCT system $\Delta\lambda$ corresponds to the FWHM of the spectrum and not to the total bandwidth. The axial resolution is specified in tissue assuming a refractive index of $n = 1.38$ for all systems.

First, Zhang et al. demonstrated functional photoreceptor imaging using a spectral-domain system [202]. The system reached A-scan rates of 1 MHz, allowing functional evaluation of phase-stable OCT volumes with a rate of up to 10 Hz. Additionally, they were able to further increase the temporal resolution of the functional changes by only evaluating B-scans with a rate of 3 kHz [202].

Besides this spectrometer-based scanning system, Azimipour et al. demonstrated the functional phase changes with a scanning swept-source system. In addition to the demands on the scanners, the light source has to be tuned fast and sufficiently stable. In this case, this was realized with a Fourier-domain mode-locked (FDML) laser with a central wavelength of $\lambda_0 = 1060\,\text{nm}$ and a bandwidth of $\Delta\lambda = 80\,\text{nm}$, which can guarantee A-scan rates of 1.64 MHz. The volume rates used for this system ranged from 6 Hz to 32 Hz [65, 199]. The specifications of both systems are also listed in Table 5.1.

Advantages of scanning systems are the better image quality and higher resolution that can be achieved with such systems compared to FF-SS-OCT. Since with scanning systems, a confocal gating is provided, multiple scattered photons that lead to additional noise and artifacts in FF-SS-OCT imaging, are suppressed. In addition, the lateral resolution is in principle a factor $\sqrt{2}$ higher with scanning OCT systems compared to FF-OCT, due to the confocal gating [203].

In order to resolve single photoreceptor cells, both systems use adaptive optics (AO) in order to correct for ocular aberrations. This allows to assign the functional changes to individual cones.

Azimipour et al. furthermore added an additional scanning light ophthalmosope (SLO) in their swept-source based OCT system, which had due to a shorter wavelength and a sub-Airy disk pinhole, an even higher lateral resolution than their AO-OCT system, to identify the position of single rods in the volume [65]. This enabled the investigation of the functional response of rods and cones independently.

Line-Field OCT An alternative technology for phase-sensitive OCT is line-field spectral-domain OCT (LF-SD-OCT). Here, instead of the entire field-of-view, only a line is imaged simultaneously, and the depth information is obtained with a spectrometer. Pandiyan et al. performed the detection with a high-speed camera (Photron, FASTCAM Mini Ax200), comparable to the one used in this thesis, where one dimension records the spatial information and the second the spectral information. By one-dimensional scanning of the retina, entire volumes can be generated. Therefore, the technique required both, a fast galvanometric scanner as well as a high-speed area camera. Requirements on the scanner, however, are lower than for the scanning systems, since only one axis needs to be scanned.

By minimizing the scanning dimension of the volume, high volume rates can be created with this technique. By doing so, Pandiyan et al. were able to detect functional signals in the photoreceptors with volume rates of up to 324 Hz utilizing their LF-SD-OCT [201]. This corresponds approximately to a doubling of the maximum time resolution, compared to our FF-SS-OCT system. By only evaluating B-scans even temporal resolution of 16 kHz were reached and thus, fast processes can be studied more precisely. In addition, the setup was combined with an AO setup and a line scanning ophthalmoscope (LSO) to improve the image quality. The specifications of the line-field setup are listed in Table 5.1.

5.1.2 Comparison of the Functional Measurements

All systems described above were able to observe an increase in the optical path length of the photoreceptor OS after optical stimulation by evaluating the phase. The general behavior of the functional change and its dependencies were very consistent across the different research groups and different OCT techniques. In the following, first, the results measured by several of these research groups are summarized and in the last part of this section, some results are described, which were so far only shown by single research groups.

Common Results In general the time course of the functional phase change reported by the different research groups corresponds to the results shown in this thesis. All measured a rapid increase of several hundred nanometers in the optical path length, which then flattens out after about 1–2 seconds. Furthermore, all research groups showed that this functional change can be attributed to single photoreceptor cells and thus the observed functional change is attributable to individual cells.

Furthermore, the dependence of the functional signal on the stimulus intensity was investigated by Zhang et al. using the scanning SD-OCT system [202] and by Pandiyan et al. using the line-field OCT system [201]. Both groups investigated the maximum optical path length change and the initial slope of the functional signal in dependence on the intensity. Their experimental data on the maximum optical path length changes corroborate the results shown in Section 4.1.1.4. Thus, with increased stimulus energy there is an increase in optical path length change in the photoreceptors OS. They could also observe a saturation for higher energies.

Likewise, both research groups observed a dependence of the initial slope on the stimulation intensity, which goes into saturation for higher intensities. In the first phase sensitive studies with FF-SS-OCT in 2016, an intensity dependence of the functional changes was also investigated. However, no dependence could be observed at that time [190]. Nevertheless, these results do not contradict the results of the other research groups or the results shown here, because the used stimulation was in the upper photopic range of vision and therefore it can be assumed that cones were in saturation.

Furthermore, Azimipour et al. also separated the contributions of rods and cones [65]. However, Azimipour et al. did not fit a curve to the measured data that was composed of two impulse responses describing the response of the rods or cones, as it was done in this thesis. Instead, their OCT images were combined with additional SLO images, which guarantee an improved lateral resolution, enabling them to resolve single rods and assign the phase changes directly to single rods and cones. This allows a clearer and more distinct differentiation of the two signal components, which leaves less room for error compared to the results shown in this thesis (Sec. 4.1.1.6).

However, the qualitative time course measured for rods and cones by Azimipour et al. corresponds in many aspects to the model used for the response of rods and cones in Section 4.1.1.6. The response in rods lasts much longer than the response in cones for the same stimulus. In addition, their measurements show that the amplitude of the cone signal is much more sensitive to the stimulus intensity when compared to rods.

These results demonstrate that phase-sensitive OCT is not limited to a specific technique, but can be applied to a wide variety of OCT systems. However, due to their different technical implementation, the different OCT systems offer the possibility to examine certain aspects of the functional signal with different levels of precision. For example, the maximum achieved temporal resolution, lateral resolution, measurement duration, and the amount of background stimulation of the measurements differ between the different systems.

Measurement Duration Among the published results, FF-SS-OCT has the longest measurement duration for functional evaluation so far, which allows the investigation of the long-term behavior of the functional signal.

Although Zhang et al. [202] and Pandiyan et al. [201] investigated the maximum optical path length increase for different stimulation intensities, they could not observe that the maximum optical path length change for high energies starts to decrease again after about 2 s of continuous stimulation and thus falls below the optical path length changes caused by lower stimulation energies (see Fig. 4.7 a). One reason, why this decrease in optical path length was not observed, may be due to the limited measurement time of 1.1 s after stimulus onset for LF-SD-OCT [201] and 2.5 s after stimulus onset for scanning SD-OCT [202]. Thus, the decrease in optical path length, which starts only after a few seconds, could not or not sufficiently be observed. The long measurement period may also be the reason why, until now, functional changes of the phase in the IPL have only been observed with FF-SS-OCT. Although the signals in the IPL seem to appear immediately after stimulation, the SNR is significantly worse, and the signal needs to reach a certain strength before it can be clearly identified as such. However, other effects might contribute to the lack of IPL signals, as well.

Nevertheless, in the case of the FF-SS-OCT, the long measurement duration goes along with a poor temporal resolution, due to the limited memory chip of the camera. Thus, the temporal resolution of 125 ms used in this thesis corresponds to the worst temporal resolution of the phase-sensitive OCT techniques discussed here.

Temporal Resolution A higher temporal resolution, however, is advantageous for investigating faster changes that last only a few milliseconds. For example, with the LF-SD-OCT system [201], the SD-OCT system [202], and the first measurements with FF-SS-OCT [190], a contraction by a few nanometers of the optical path length of the photoreceptor OS was observed for only a few milliseconds before the strong increase shown in this thesis started. While the temporal resolution of the FF-SS-OCT system (6 ms) was sufficient to observe this contraction qualitatively, the other groups were able to increase the temporal resolution even further by using only B-scans instead of complete volumes for their evaluation. Zhang et al. were able to achieve temporal resolutions of 0.33 ms with their SD-OCT systems. The highest temporal resolution, however, was achieved with the LF-SD-OCT system with significantly higher B-scan rates of 16 kHz. This allowed a quantitative analysis of the initial contraction, which in turn allowed the LF-SD-OCT system to determine the dependence of the stimulus intensity on the slope and the amount of the contraction [201].

Single Cell Analysis The two scanning OCT systems are superior to the other systems in terms of image quality and lateral resolution. Accordingly, those two systems have increasingly focused on single-cell analyses of the functional signal. Zhang et al. for example have demonstrated that a single-cell analysis of the functional change in dependence on different spectral stimuli allows the spectral classification of the individual cones into S-, M- and L-cones [202].

As already mentioned above, the increased lateral resolution of Azimipour et al. , due to the additional SLO system, even allows resolving the functional changes for single rods, and thus allows distinguishing the functional contributions of rods and cones [65]. In addition to the high lateral resolution of Azimipour et al., the long central wavelength of 1060 nm results in a much lower background stimulation caused by the OCT measurement itself, compared to the other systems. This allows examination of the functional changes at significantly lower stimulation intensities, which is particularly interesting for the investigation of rods. By switching to weaker stimulation intensities in the middle mesopic range (down to a photopigment bleaching of 0.007% in rods), Azimipour et al. were able to show that the initial slope of the functional signal in the rods is also dependent on the stimulation intensity when using intensities in the mesopic range. The behavior is analogous to that of the cone signal. This means, that in a certain intensity range the initial slope increases with stimulation intensity, while there is a saturation for higher intensities. They observed the saturation at approximately 80 nm/s, which is exactly the slope that is measured for the rod signal in the model of Section 4.1.1.6.

5.2 Origin of the Observed Phase Change

One drawbacks of functional retinal imaging with phase-sensitive OCT is, that it is not known what physiological processes are responsible for the observed changes in the optical path length. In general, there are two different parameters, which can alter the optical path length of a layer. First, it changes if the observed layer thickness l changes, and secondly, it changes if the the refractive index of the layer changes.

The contributions of the physical length and the refractive index to the change in the optical path length ΔOPL can be broken down as follows.

$$\Delta\mathrm{OPL} = (n+\Delta n)(l+\Delta l) - nl \tag{5.1}$$

$$= nl + l\Delta n + n\Delta l + \Delta n\Delta l - nl \tag{5.2}$$

$$\simeq l\Delta n + n\Delta l \tag{5.3}$$

Here n represents the refractive index and l the physical length of the investigated layer. The last approximation is based on the assumption that the changes Δn and Δl are small relative to the initial values of the refractive index and the length, so that the term $\Delta l \Delta n$ can be neglected.

A variety of different effects, ranging from thermal expansion, via neuro-vascular coupling and concentration changes to changes in the volume of the cell or a mechanical conformation change could lead to both, a change in the layer thickness or a change in the refractive index. Based on the observed behavior of the phase change, different possible origins are discussed in the following.

5.2.1 Photo Damage or Thermal Expansion

In order to prove the physiological relevance of the functional phase change, the first step is to make sure that they are not caused by non-physiological processes like photo-damage or tissue heating caused by the stimulation light. The physical length can change due to the thermal expansion of the tissue. The expansion of the tissue in turn reduces the mass density, which in turn would reduce the refractive index. Therefore, tissue heating and photo-damage could alter the optical path length via the refractive index as well as by the physical length.

However, for the signals in the IPL tissue heating and photo-damage can be easily ruled out, since the changes do not occur at the lateral positions stimulated with light. This is not the case for the changes in the photoreceptors, where the extension of the optical path length occurs exactly in the areas stimulated with light. In addition, the changes of several hundred nanometers, by which the optical path length increases are large for cellular processes. The maximum used stimulus luminance of $4600\,\mathrm{cd/m^2}$ is perceived as extremely bright by the dark-adapted volunteer since it causes for a dark-adapted eye a complete saturation of the photocurrent. Although it is perceived as extremely bright, the actual energy deposited in the retina is far below the maximum permissible exposure. In fact, the luminance approximately corresponds to the luminance the retina is exposed to under normal daylight conditions ($8000\,\mathrm{cd/m^2}$ average clear sky). It is therefore not to be assumed that photo-damage in the retina occurs under such light conditions. Furthermore, the phase changes are not permanent but are fully reversible for short stimuli, which also contradicts photodamage as a possible origin for the observed change.

A thermal expansion can also be excluded, since the increase in the optical path length should stop at the moment the energy deposition in the retina stops. However, with a short stimulus of 100 ms duration the increase in the optical path length continues long after the stimulus ends (Sec. 4.1.1.6).

5.2.2 Neuro-Vascular Coupling and Increased Metabolic Demands

Another explanation for these huge changes in the optical path length is that a change in blood flow is caused by an increased metabolic demand of the neurons. Such a change in blood flow could lead to a change in the vessel diameter, which in turn could lead to a change of the layer thickness of the surrounding tissue. Furthermore, a change in the blood oxygen level due to higher demand could lead to a change in the refractive index as well.

While the upper neuronal layers are supplied by the vessel system emerging from the optic nerve, the photoreceptor cells are supplied by the underlying RPE, which in turn is supplied by the underlying choroid. Thus, in the close proximity of the photoreceptor cells, there are no vessels that could alter the optical path length by a change in blood flow. Besides, in Section 4.1.1.3 it could be demonstrated, that the functional change in the photoreceptor layer could be assigned to single photoreceptor cells, which strongly suggests that the changes have a cellular origin and are not a byproduct of the neuro-vascular coupling.

For the signals in the IPL, the neuro-vascular coupling cannot be completely excluded as a possible cause, since the signals in this level could not be assigned to individual cells. Additionally, in the IPL the finer capillary system is stratifying, while the bigger vessels are stratifying in the ganglion cell layer. This means, that in general a much finer functional change can be created by neuro-vascular coupling in the IPL than could be expected in the cellular level. However, there are some indications that make neuro-vascular coupling unlikely as a cause. First of all, the observed functional changes in the IPL do not show any correlation with the capillary system. Second, the fact that the sensitivities of the functional signals differ in different levels of the IPL and even show opposite sign also contradicts neuro-vascular coupling as a cause for the observed functional changes.

5.2.3 Concentration Changes

During the activation of neurons several molecular path ways are initiated or shut down, during which the concentration of the participating molecules can change. Such concentration changes can change the refractive index of a cell. Here, different molecules can have varying degrees of influence on the refractive index. For example, changes in the ion concentration cause significantly higher effects than changes in protein concentration [204]. At wavelengths in the visible and NIR, a linear increase in the refractive index of $\frac{\Delta n}{\Delta c} = 0.01\,\mathrm{M}^{-1}$ with an increase in ion concentration is observed [204]. In other words, an increase of the particle number N in the cell by ΔN leads to a refractive index change of

$$\Delta n = 0.01\,\mathrm{L/mol}\frac{\Delta N}{V}, \tag{5.4}$$

where V describes the volume of the cell. But, a change in concentration also creates an osmotic pressure, which leads to an influx of water into the cell until the physiological concentration of the cell is restored. According to this equation, the following relationship for the volume change applies:

$$c = \frac{N_1}{V_1} = \frac{N_1 + \Delta N}{V_1 + \Delta V}. \tag{5.5}$$

Here $N_1 + \Delta N$ and $V_1 + \Delta V$ describes the new number of particle and the new volume, respectively, and N_1 and V_1 the corresponding values for the old volume. From Equation (5.5) the following relation for the volume change ΔV can be derived:

$$\Delta V = \Delta N \frac{V_1}{N_1} = \frac{\Delta N}{c_1}. \tag{5.6}$$

Such a change in volume can lead to a change in length of the cell, which also influences the optical path length.

The following consideration shall help to estimate the magnitude of the influence of these two parameters, i.e., the impact of the volume change and the refractive index change on the optical path length. For this purpose, the volume of a cylinder $V = Al$ is assumed, where A represents the base area and l is the length of the cylinder.

The optical path length change due to the change of the refractive index, $\Delta\mathrm{OPL}_n$, is according to Equation (5.4) given by

$$\Delta\text{OPL}_n = \Delta n l = 0.01\,\text{L/mol}\frac{\Delta N}{V}l = \frac{l\Delta c}{100\,\text{mol/L}}. \tag{5.7}$$

In order to evaluate the influence that an osmotic expansion would have on the optical path length, it is first assumed that the volume change is only compensated by a change in the length l of the cylinder. Accordingly, ΔOPL_l can be described as

$$\Delta\text{OPL}_l = n\Delta l = n\frac{\Delta V}{A} = nl\frac{\Delta c}{c_1}, \tag{5.8}$$

where for the last equality the relation from Equation (5.6) is used. The ratio of the two optical path length changes, ΔOPL_n and ΔOPL_l is thus

$$\frac{\Delta\text{OPL}_l}{\Delta\text{OPL}_n} = n\frac{100\,\text{mol/L}}{c_1}. \tag{5.9}$$

For the physiological osmolarity of photoreceptors (325 mOsm [205]), this means that when the concentration changes, the resulting length changes have a 300 times greater effect on the optical path length change than the corresponding changes in the refractive index. Even if the volume change is compensated by an expansion in all three spatial dimensions, ΔOPL_l is still greater than ΔOPL_n by a factor of 100 under the condition that $l \gg \Delta l$.

It can therefore be assumed that refractive index changes in the cell have a negligible effect on the signals observed in this thesis. In addition, such changes are only of short duration because they are directly compensated by an osmotically driven water flow, which also does not fit to the signals observed here, which last for several seconds. A volume change due to a concentration changes, on the other hand, is still a possible origin for the observed signal.

5.2.4 Causes for Osmotical Change

In order to estimate how big osmotical volume changes could be under physiological condition, several concentration changes, which occur during the activation of photoreceptor cells and their maximum influence on the photoreceptor length are discussed in the following section, which means that it is assumed that the volume change is compensated only in one dimension parallel to the $z-$axis. Thus, for a simple cylindrical volume, the relative change in concentration $\Delta r_c = {}^{\Delta c}/_{c_1}$ corresponds to the relative change in length $\Delta r_l = {}^{\Delta l}/_{l}$. However, about 50% of the rod's OS volume is occupied by the discs, which means that in this case $\Delta r_c = 2\Delta r_l$ applies. Therefore, the optical path length change can be described by $\Delta\text{OPL}_l = 0.5n\Delta r_c l$.

For several possible osmolytic active molecules that occur during phototransduction, the corresponding value of concentration change Δc, relative concentration change Δr_c regarding the physiological osmolarity of the cell (325 mOsM) [205] and the resulting maximum length change Δl_{max} of the OS are listed in Table 5.2. For this, a starting length of $l = 25\,\mu\mathrm{m}$ [102] was assumed for the OS.

The first step of the phototransduction, the activation of rhodopsin molecules, was already excluded in previous studies with knock-out mice as a cause of the observed elongation of the photoreceptor OS [206]. Therefore the next step of the phototransduction is the transducin activation. During the activation, these molecules are divided into two sub-units, one of which acts as an additionally osmolytically active particle in the cytoplasm. However, a complete activation of all transducin molecules can only cause a maximum concentration change of $\Delta c = 300\,\mu\mathrm{M}$ [101]. The change in concentration thus achieved corresponds to about 0.1% of the physiological concentration of the cell. Therefore, this should result in a volume increase of 0.1%, which leads to a maximum length change of 16.4 nm under the above mentioned conditions. That change in the OS length is about one order of magnitude smaller than the changes in the optical path length observed in this thesis. Therefore, the concentration change in the tranducin molecules can be excluded as cause for the functional change.

The next molecule activated in the phototransduction cascade is PDE. When PDE is activated, no new sub-units are split off, so there is no change in concentration in the cell. The activated PDE hydrolysis cGMP molecules to GMP molecules. However, this hydrolysis does not change the osmolarity, either. Although, the GMP concentration is increased by this step, the cGMP concentration is reduced to the same extent, so that there is no change in the osmolytically active particles. Also the cGMP molecules released by the closing CNG-channels correspond to only an increase of 1 µOsM [114], when all channels are completely closed and are therefore also negligible as a cause.

The last step of phototransduction is the photocurrent. Here, the maximum photocurrent across the membrane is 13 pA [116]. The electrical charge of the photocurrent is carried by 80% single charge sodium ions and by 20% double charged ions (15% magnesium, 5% calcium) [114]. This means that the concentration change Δc caused by the photocurrent I_{PC} can be calculated with the help of the elementary charge $e = 1.602 \times 10^{-19}$ C and the correction factors ($r_{\mathrm{Na}} = 0.8$; $r_{\mathrm{Ca}} = 0.025$ and $r_{\mathrm{Mg}} = 0.075$) according to:

$$\Delta c = \frac{I_{\mathrm{PC}}}{e \times (r_{\mathrm{Na}} + r_{\mathrm{Ca}} + r_{\mathrm{Mg}}) \times V \times N_{\mathrm{A}}} = 3.4\,\mathrm{mOsM/s}$$

Here, $V = 40\,\mu\mathrm{m}^3$ represents the volume of the OS solvent [102] and $N_A = 6.022 \times 10^{23}$ is the Avogadro constant. This change in concentration approximately correspond to 1% of the physiological osmolarity of the cell. If one assumes again a purely axial change of

Molecules	Max. Δc [µM]	Δr_c [%]	$\Delta l_{\max}$ [nm]
Transducin	300	0.1	16.4
PDE	0.0	0.0	0.0
cGMP/GMP	0.25	0.01	1.5
Photocurrent	−3000	−1	−164

Table 5.2: Concentration changes of different molecules during the phototransduction: Max. Δc is the maximum concentration change of the respective molecules that can be reached in rods, Δr_c [%] is the maximum relative change in concentration the molecule can cause regarding the physiological osmolarity of the cell, $\Delta l_{\max}$ [nm] is the maximum length change that can be reached by the concentration changes considering a purely axial expansion of the volume.

the OS, then such a concentration change results in a length change of 164 nm. Which is in the same order of magnitude, than the observed functional changes in the optical path length of the photoreceptor OS.

However, the photocurrent does not lead to an accumulation of the ions in the cell but in the extracellular matrix around the photoreceptors. This reduces the concentration in the cell and as a result the volume of the OS should decrease. If the observation of the phase signal is really due to osmotic volume increase, the sought reaction must not only increase the cellular concentration by a few mOsM but also compensate for the decrease in concentration caused by the photocurrent. Yet, concentration changes of this magnitude are not very likely, as evident from the other steps of the phototransduction.

In fact, there are several experiments that have measured the volume change of the extracellular matrix in the retina during light stimulation. These measurements show that the volume around the photoreceptor OS increases with light stimulation [207–209]. These volume changes are highly correlated with changes in the concentration of sodium and calcium ions, which are assumed to be caused by the photocurrent and thus should correlate with a reduction in the volume of the OS.

However, the previous considerations were based on the assumption that the volume change is caused by a purely axial expansion of the photoreceptor. This is equivalent to a corresponding expansion of the plasma membrane. Yet, plasma membranes are very incompressible [210] and tend to rupture after only a few percent stretching [211]. It is known from various cells such as blood cells that they change their conformation to adapt to different cell volumes without changing the size of their membrane surface [212, 213]. Thereby, a variety of different conformation changes are possible for the photoreceptor cells in order to change their volume while maintaining membrane surface. Even conformation changes are conceivable that would combine a volume reduction with an elongation of the OS.

The simplest model would be to describe the photoreceptor OS as a biaxial ellipsoid with a short axis a and a long axis b. Assuming that the surface of this ellipsoid, i.e., the plasma membrane area, is constant, the maximum volume for the cell is reached when $a = b$, i.e., when the OS is completely spherical. Such a behavior was indeed observed for photoreceptor OS under extreme hypotonic conditions [214]. In contrast, the volume becomes smaller and smaller as the radii a and b diverge. That is, assuming a constant surface area, an extension of the photoreceptors is compatible with a volume reduction caused by a confrontational change.

5.2.5 Potential Changes

The change in the ion concentration not only causes a volume change of the cell, but also changes the local electric field across the plasma membrane. These changes can alter the membrane properties, which can in turn alter the conformation of the cell's plasma membrane. Those deformation of the plasma membrane can alter the length of a cell, and therefore are a possible origin for the phase-changes.

Such deformation of the cell membrane has already been observed for different cell types, such as crustaceous nerves, squid axons, mammalian nerves or HEK293 cells, during action potentials [59, 61, 215–217]. However, mostly these deformation are only in the order of a few nanometers during the action potentials. These effects are therefore much too small to explain the effects observed here. However, there is the possibility that such an effect can accumulate over the length of the cone OS due to the special structure of the cone OS, which consists of the tightly stacked membrane disks, achieving greater effects in cones.

5.3 Correlation of the Phase Changes

The two possible mechanisms for the increase in the optical path length discussed in Sections 5.2.4 and 5.2.5 are both based on the photo current. And indeed, there are certain correlations between the functional changes observed in this thesis and the photo current.

First, the slope of the functional signal in the photoreceptors changes with the stimulation intensity over a range of 2–3 orders of magnitude, as does the photocurrent, which then also goes into saturation for higher intensities [116]. The observed functional change in the photoreceptor cells also lasts for several seconds, just like the photocurrent, and also shows a reduction of the signal strength after a few seconds of continuous stimulation due to adaptation processes [218].

There are also some correlations with the potential change in the IPL and the signals observed there. On the one hand, the bipolar cells generate a generator potential, just as the photoreceptors do, which means that the induced current in the upper cell levels lasts for several seconds, too, and their amplitude can be scaled according to the visual stimulus, which also fits with the long lasting phase changes in the IPL.

In addition, ON and OFF cells react with mutually inverse potential changes for the same visual stimulus. This is also consistent with the phase changes observed in the IPL, where for the same stimulus, an increase in optical path length is observed for the levels where the synaptic connections of the ON bipolar and ganglion cells are located, while a decrease of the optical path length is observed in the layers where the synaptic connections of the OFF bipolar and ganglion cells are located.

Although the kinetics of the optical path length changes and the photocurrent fit well, there are still other mechanisms possible. For example, the conformation of the cells could also change by an active process of the cytoskeleton [219]. Furthermore, since the optical path length can be influenced by a huge variety of molecular processes it is possible, that the changes in the photoreceptor layer and in the IPL are caused by very different molecular processes.

6 Conclusion

Cellular imaging of neuronal cells has great potential for the clinical diagnosis of retinal and neurological diseases. In addition to morphological changes, functional changes are of special interest as they may provide the possibility for early diagnosis, in particular for degenerative diseases. Therefore, different imaging modalities have addressed functional retinal imaging in the past.

In this thesis, an imaging modality was presented that allows functional contrast in the retina by using the phase information of the OCT signal to detect minute changes in the optical path lengths of retinal layers. Such changes can have different molecular causes and therefore are not restricted to certain cell classes but may be present in all activated areas of the retina.

In addition, phase imaging with OCT allows the spatial localization of the functional signal in all three spatial directions with $5 - 10\,\mu m$ resolution, which allows a differentiated investigation of the functional signals in the different retinal structures. For example, the high lateral resolution of the functional signals enables assigning the functional signals to individual photoreceptors.

Additionally, the investigation of the lateral information processing of higher neurons becomes possible, which was for example demonstrated by observing an increased signal at the edges of the visual stimulus. Thus, phase-sensitive OCT currently offers a more detailed analysis of retinal functionality than other methods which can only detect an integrated signal of retinal activity.

Furthermore, the axial resolution of the phase-sensitive OCT makes it possible to measure different neuronal layers of the retina simultaneously, which enables, for the first time, studying the functional interconnection of photoreceptors and ganglion cells in living humans. It thereby offers the possibility to investigate the information processing and inter-neural communication of the retina in more detail. The axial differentiation of functional signals in the IPL even allows distinguishing the functional contributions of different bipolar/ganglion cell types. Thus, different sensitivities of individual layers of the IPL to different visual parameters can be investigated by this functional contrast, allowing researchers to study the classification of the human ganglion and bipolar cell types for the first time in the living human eye.

The functional contrast provided by phase-sensitive OCT can therefore become important for physiological research and clinical diagnosis. Consequently, it would be desirable to establish this functional contrast in clinical routine. The chances for a clinical transfer of this technique are very good since the OCT technique has already been established as a standard technique in ophthalmology.

The only prerequisite for such a functional evaluation of the phase information is a sufficiently phase-stable acquisition of the OCT data. In this thesis, this was guaranteed by an axial and lateral parallelization of the data with an FF-SS-OCT system. However, phase-sensitive functional OCT has been demonstrated with several other OCT techniques and, thus, there may be several ways to bring the demonstrated functional contrast into the clinics.

7 Outlook and Future Work

The functional evaluation of retinal neurons using phase-sensitive OCT is a very young field of research. Although the potential of this technique has been impressively demonstrated in recent years, its full extent will only be revealed in the coming years. For the continuation of this research, there are different worthwhile directions. On the one hand, it is certainly important to further develop the FF-SS-OCT technology to achieve a better image quality and a more precise evaluation of the functional signals. On the other hand, the current FF-SS-OCT technology has little chance of being transferred to the clinics due to its expensive high-speed area camera and the lack of live imaging capability. Therefore, new technological solutions need to be investigated. Finally, the functional signals have to be further characterized in order to identify the molecular cause and to evaluate the clinical relevance of the signals.

7.1 Further Developments of FF-SS-OCT

FF-SS-OCT enables phase-locked imaging, which is of great advantage in the field of computational optics and functional imaging as shown in this thesis. However, FF-SS-OCT has its drawbacks. In comparison, both, resolution and image quality of FF-SS-OCT are worse than in scanning OCT systems.

For example, the axial resolution of swept-source OCT is constrained by the available tuneable light sources. Currently, usable lasers, that do not exhibit significant amplitude and phase noise and provide sufficient imaging depth, are limited to 100 nm bandwidth or less due to the available gain media. Therefore, a maximum axial resolution of 7 µm in air is achieved by swept-source systems. Spectrometer-based systems, on the other hand, can achieve much larger bandwidths and thus higher axial resolution.

A higher axial resolution is of particular interest for functional imaging in the IPL. Here, a better understanding of the different ganglion and bipolar cell types can be achieved if an individual evaluation of each of the 5 strata can be guaranteed. Yet, due to the current axial resolution of FF-SS-OCT, only 3 layers can be evaluated.

One possibility, which could be explored in the future to reach a higher axial resolution for FF-SS-OCT despite the restrictions, is to combine multiple swept sources with different but overlapping spectral ranges to increase the spectral bandwidth.

Another main disadvantage is that the area illumination of the sample does not allow confocal gating for FF-SS-OCT imaging. The amplitude transfer function of imaging modalities using confocal gating covers twice the frequencies of those that do not. Thus an up to $\sqrt{2}$–fold improvement of lateral resolution can be achieved by using a confocal gating [203].

Another advantage achieved by confocal gating is that multiple scattered photons are suppressed. Due to the multiple scattering in the tissue, the photons are erroneously assigned to deeper levels, which leads to an additional noise background in the neuronal layers for FF-SS-OCT. Furthermore, imaging below the RPE is no longer possible due to the high concentrations of the strong scattering melanin granules in the RPE cells.

Still, improving lateral resolution and image quality will provide additional useful information for functional imaging. For example, while it is currently possible to resolve individual photoreceptor cells and thus perform functional imaging of individual cells at this level of the retina, this is not yet possible for ganglion and bipolar cells using FF-SS-OCT. The cellular imaging of ganglion cells in the retina using a scanning system has already been demonstrated [220], i.e., it is in general possible with OCT, but so far this resolution and image quality could not be combined with functional contrast. Assigning the functional changes to individual ganglion and bipolar cells, however, could prove very helpful for a further classification of those different cell types. Similarly, a functional evaluation in the RPE and deeper layers would be of interest, which is currently not possible due to the multiple scattered photons.

There are several ways to get around the two limitations mentioned above. To increase the lateral resolution, structured illumination of the retina or illumination from different angles (oblique illumination) can be used. By changing the illumination in this way, the lateral frequency spectrum is increased and thus the resolution improved.

Furthermore, it has recently been successfully shown that multiple scattered photons can be suppressed in FF-SS-OCT by introducing a deformable mirror with a rapidly changing surface in the illumination beam, which effectively destroys the spatial coherence of the illumination [221, 222].

7.2 Exploration of Alternative Techniques for the Clinical Transfer

Although the physiological significance of functional phase contrast using OCT can obviously be advantageous in clinical diagnostics, the current setups that can achieve such contrast are hardly suitable for clinical applications. In FF-SS-OCT, the need for a high-speed 2D camera for detection is a major obstacle. Such a camera is expensive (the FASTCAM SA-Z camera used here costs about 100,000 Euro) and does not allow a live readout of the data. The same problem arises with the line-field spectrometer-based OCT [201], which also depends on a high-speed 2D camera. In addition, the line-field setup has an integrated wavefront sensor and deformable mirror for AO to ensure cellular analysis of the functional signals. AO setups are usually expensive and complex to adjust, which is also an additional obstacle for a commercial system.

Probably the highest chances of being transferred into clinical systems have scanning OCT systems. They provide a live preview, flexible choice of the field of view, and already represent the dominant technology in commercial and approved medical OCT imaging. But for the scanning OCT systems, the additional AO or SLO systems represent an extra complexity of the setup [64, 65, 202], which could be an obstacle to the commercialization of the system. However, since for a functional phase contrast in general phase-locked measurements are needed it would be worth investigating if, after suitable phase correction, the algorithms for numerical aberration correction work for scanning OCT systems as well. Furthermore, it would be helpful to investigate the minimal required scanning speed to generate phase-locked volumes and the optimal scanning pattern to minimize motion artifacts.

7.3 Functional Imaging of the Entire Retina

As has been shown in this work, functional changes cannot only be observed in one retinal layer but in two different layers at the same time. Provided that the functional changes correlate with the potential changes in the neurons, it can be assumed that functional changes can also be detected in further retinal layers. For example, in the OPL, which contains the connections of the photoreceptors with the bipolar and horizontal cells. This layer is slightly thinner and less light is scattered back as from the IPL, which degrades the SNR of the phase contrast. Nevertheless, within the scope of this thesis, it was possible to detect functional changes in this layer using FF-SS-OCT (see Fig. 7.1). These signals are much weaker than the ones in the IPL and photoreceptor OS and therefore difficult to identify as such. For this reason, the detection of these functional changes in the OPL is so far only possible for a few measurements. It is not yet clear what causes these changes to be detectable in some measurements and not in others. However, it is noticeable that these changes are mainly visible in measurements where there is little lateral movement of the eye. Since the stimulation is static, lateral movements lead to a smearing of the stimulated area on the retina. It is therefore possible that the change in the OPL is increased by a high contrast, which is reduced by a smeared image. If this is the case, an eye-tracking system that detects eye movement and readjusts the stimulation accordingly would ensure that these signals can be reproducibly detected in all measurements.

A further, interesting retinal structure is the nerve fiber layer. Here, action potentials are transmitted to the visual cortex. The challenge when imaging action potentials is that they cause only very short potential changes. Accordingly, it can be expected that

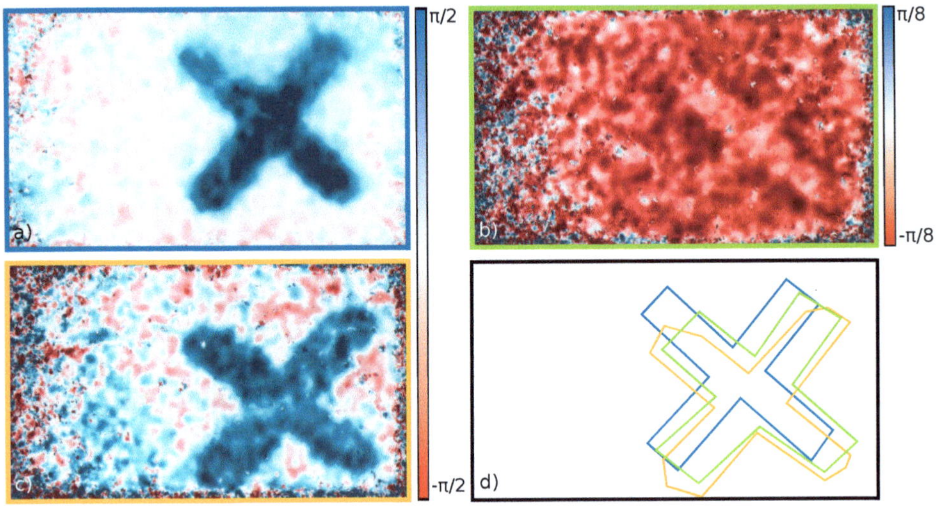

Figure 7.1: Functional signal of the full retina in the periphery, 3 mm superior the fovea. a) Functional signal in the photoreceptor OS. b) Functional signal in the OPL and, c) functional signals in the IPL. d) Lateral shift of the functional signal in the different layers. Blue displays the photoreceptor OS signal, green the signal in the OPL, and yellow the signal in the IPL.

only very short changes of optical properties will occur. Nevertheless, it has already been shown in vitro that action potentials can be made visible by phase-sensitive imaging [58, 59, 61, 217]. Furthermore, the nerve fiber layer in OCT usually offers sufficient backscattered intensity and thus a good SNR for the phase evaluation, so that even small changes could be detectable. However, the time resolution achieved for functional, phase-sensitive FF-SS-OCT is so far too low to detect the rapid changes of less than 2 ms.

Still, measuring the transmission velocity of action potentials in vivo and in humans could be of great interest from a medical point of view. It would therefore be worthwhile to look for ways to increase the temporal resolution of the phase-sensitive FF-SS-OCT.

7.4 Investigating of the Molecular Origin

To use functional imaging by phase evaluation of OCT signals clinically, it would be advantageous to be able to prove or clearly identify the molecular origin of the signals. In this way, pathological changes in the functional signals can be interpreted more directly on a molecular level.

In order to clarify the molecular origin experimentally, measurements on ex-vivo tissue or in animals will be necessary, as it enables specifically switching off or manipulating signal pathways. First preliminary investigations have already been carried out using explanted porcine retina [223, 224]. In these preliminary studies, functional signals in the photoreceptors could be measured by FF-SS-OCT. However, the results were too unreliable to study the signaling pathways. Therefore, the next step will be to stabilize the experimental parameters to ensure reproducible time courses of the functional changes in the explanted retina before molecular manipulation of different signal pathways can be performed.

7.5 Clinical Studies

In addition to the ex-vivo experiments, it is also important to correlate changes in the functional signals with specific pathologies. It is conceivable that the first preliminary clinical investigations can be carried out on the FF-SS-OCT laboratory setup, for example, to investigate the changes in the functional signals for degenerative diseases of the retina, such as glaucoma.

Following, larger clinical studies will have to be conducted to quantify the clinical relevance of phase-based functional contrast. If one wants to investigate the relevance for the early diagnosis of neurodegenerative diseases of the CNS, even long-term studies may have to be carried out. However, it should be remembered that in an increasingly aging society, the number of such diseases will continue to rise. According to the German Center for Neurodegenerative Diseases (DZNK), the number of patients suffering from neurodegenerative diseases will rise to 3 million in Germany alone by the year 2050. So far, none of those diseases are curable. However, once they are diagnosed they are in many cases treatable and their progression can be slowed down. An imaging system, which can identify the diseases before the health of the patient is significantly affected could thereby improve the diagnosis and treatment of an increasing number of patients.

Due to its high sensitivity and the possibility to differentiate the contributions of different cell classes or cell types, the functional phase-contrast demonstrated in this work has a great potential to identify pathological abnormalities of retinal neuronal function at an early stage. It is therefore promising that phase-sensitive OCT may be used in the future for the early diagnosis of neurodegenerative diseases and help to identify such diseases at a stage before the patient's quality of life is already impaired.

Bibliography

[1] A. London, I. Benhar, and M. Schwartz, "The retina as a window to the brain — from eye research to cns disorders," *Nature Reviews Neurology*, vol. 9, no. 1, pp. 44–53, 2013.

[2] J. W. Streilein, "Ocular immune privilege: Therapeutic opportunities from an experiment of nature," *Nature Reviews Immunology*, vol. 3, no. 11, pp. 879–889, 2003.

[3] C. Kaur and E. A. Ling, "Blood brain barrier in hypoxic-ischemic conditions," *Current Neurovascular Research*, vol. 5, no. 1, 2008.

[4] W. I. McDonald and D. Barnes, "The ocular manifestations of multiple sclerosis. 1. Abnormalities of the afferent visual system.," *Jounal of Neurology, Neurosurgery and Psychatry*, vol. 55, pp. 747–752, 1992.

[5] U. Leibowitz and M. Alter, "Optic nerve involvement and diplopia as initial manifestations of multiple sclerosis," *Acta Neurologica Scandinavica*, vol. 44, no. 1, pp. 70–80, 1968.

[6] T. L. Sørensen, J. L. Frederiksen, H. Brønnum-Hansen, and H. C. Petersen, "Optic neuritis as onset manifestation of multiple sclerosis," *Neurology*, vol. 53, no. 3, pp. 473–481, 1999.

[7] A. J. Green, S. McQuaid, S. L. Hauser, I. V. Allen, and R. Lyness, "Ocular pathology in multiple sclerosis: Retinal atrophy and inflammation irrespective of disease duration," *Brain*, vol. 133, no. 6, pp. 1591–1601, 2010.

[8] J. B. Fisher, D. A. Jacobs, C. E. Markowitz, S. L. Galetta, N. J. Volpe, M. L. Nano-Schiavi, M. L. Baier, E. M. Frohman, H. Winslow, T. C. Frohman, P. A. Calabresi, M. G. Maguire, G. R. Cutter, and L. J. Balcer, "Relation of visual function to retinal nerve fiber layer thickness in multiple sclerosis," *Ophthalmology*, vol. 113, no. 2, pp. 324–332, 2006.

[9] M. L. Monteiro, D. B. Fernandes, S. L. Apóstolos-Pereira, and D. Callegaro, "Quantification of retinal neural loss in patients with neuromyelitis optica and multiple sclerosis with or without optic neuritis using Fourier-domain optical coherence tomography," *Investigative Ophthalmology & Visual Science*, vol. 53, no. 7, pp. 3959–3966, 2012.

[10] Ö. Altintaç, P. Içeri, B. Özkan, and Y. Çağlar, "Correlation between retinal morphological and functional findings and clinical severity in Parkinson's disease," *Documenta Ophthalmologica*, vol. 116, no. 2, pp. 137–146, 2008.

[11] D. Devos, M. Tir, C. A. Maurage, N. Waucquier, L. Defebvre, S. Defoort-Dhellemmes, and A. Destée, "ERG and anatomical abnormalities suggesting retinopathy in dementia with Lewy bodies," *Neurology*, vol. 65, no. 7, pp. 1107–1110, 2005.

[12] N. K. Archibald, M. P. Clarke, U. P. Mosimann, and D. J. Burn, "The retina in Parkinson's disease," *Brain*, vol. 132, no. 5, pp. 1128–1145, 2009.

[13] H. V. Danesh-Meyer, H. Birch, J. Y. Ku, S. Carroll, and G. Gamble, "Reduction of optic nerve fibers in patients with Alzheimer disease identified by laser imaging," *Neurology*, vol. 67, no. 10, pp. 1852–1854, 2006.

[14] J. C. Blanks, D. R. Hinton, A. A. Sadun, and C. A. Miller, "Retinal ganglion cell degeneration in Alzheimer's disease," *Brain Research*, vol. 501, no. 2, pp. 364–372, 1989.

[15] J. C. Blanks, Y. Torigoe, D. R. Hinton, and R. H. Blanks, "Retinal pathology in Alzheimer's disease. I. Ganglion cell loss in foveal/parafoveal retina," *Neurobiology of Aging*, vol. 17, no. 3, pp. 377–384, 1996.

[16] V. Parisi, R. Restuccia, F. Fattapposta, C. Mina, M. G. Bucci, and F. Pierelli, "Morphological and functional retinal impairment in Alzheimer's disease patients," *Clinical Neurophysiology*, vol. 112, no. 10, pp. 1860–1867, 2001.

[17] A. Guevara-Torres, A. Joseph, and J. B. Schallek, "Label free measurement of retinal blood cell flux, velocity, hematocrit and capillary width in the living mouse eye," *Biomedical Optics Express*, vol. 7, no. 10, p. 4228, 2016.

[18] J. Tam and A. Roorda, "Speed quantification and tracking of moving objects in adaptive optics scanning laser ophthalmoscopy," *Journal of Biomedical Optics*, vol. 16, no. 3, p. 036002, 2011.

[19] J. A. Martin and A. Roorda, "Direct and noninvasive assessment of parafoveal capillary leukocyte velocity," *Ophthalmology*, 2005.

[20] Z. Zhong, G. Huang, T. Y. P. Chui, B. L. Petrig, and S. A. Burns, "Local flicker stimulation evokes local retinal blood velocity changes," *Journal of Vision*, vol. 12, 2012.

[21] R. A. Leitgeb, R. M. Werkmeister, C. Blatter, and L. Schmetterer, "Doppler optical coherence tomography," *Progress in Retinal and Eye Research*, vol. 41, pp. 26–43, 2014.

[22] A. S. G. Singh, C. Kolbitsch, T. Schmoll, and R. A. Leitgeb, "Stable absolute flow estimation with Doppler OCT based on virtual circumpapillary scans," *Biomedical Optics Express*, vol. 1, no. 4, p. 1047, 2010.

[23] G. T. Feke and C. E. Riva, "Laser Doppler measurements of blood velocity in human retinal vessels," *J. Opt. Soc. Am.*, vol. 68, no. 4, pp. 526–531, 1978.

[24] C. Riva, B. Ross, and G. B. Benedek, "Laser Doppler measurements of blood flow in capillary tubes and retinal arteries.," *Investigative Ophthalmology*, vol. 11, no. 11, pp. 936–944, 1972.

[25] L. Puyo, M. Paques, M. Fink, J.-A. Sahel, and M. Atlan, "In vivo laser Doppler holography of the human retina," *Biomedical Optics Express*, vol. 9, no. 9, pp. 4113–4129, 2018.

[26] L. Puyo, M. Paques, M. Fink, J.-A. Sahel, and M. Atlan, "Waveform analysis of human retinal and choroidal blood flow with laser Doppler holography," *Biomedical Optics Express*, vol. 10, no. 10, pp. 4942–4963, 2019.

[27] L. Puyo, M. Paques, M. Fink, J.-A. Sahel, and M. Atlan, "Choroidal vasculature imaging with laser Doppler holography," *Biomedical Optics Express*, vol. 10, no. 2, pp. 995–1012, 2019.

[28] C. E. Riva, S. Harino, R. D. Shonat, and B. L. Petrig, "Flicker evoked increase in optic nerve head blood flow in anesthetized cats," *Neuroscience Letters*, vol. 128, no. 2, pp. 291–296, 1991.

[29] V. V. Toi and C. E. Riva, "Variations of blood flow at optic nerve head induced by sinusoidal flicker stimulation in cats," *Journal of Physiology*, vol. 482, no. 1, pp. 189–202, 1994.

[30] G. G. Stokes and R. S. Lucasian, "On the reduction and oxidation of the colouring matter of the blood," vol. 8, no. 2, pp. 355–364, 1864.

[31] J. B. Hickam and R. Frayser, "Studies of the retinal circulation in man. Observations on vessel diameter, arteriovenous oxygen difference, and mean circulation time," *Circulation*, vol. 33, no. 2, pp. 302–316, 1966.

[32] D. Schweitzer, M. Hammer, J. Kraft, E. Thamm, E. Konigsdorffer, and J. Strobel, "In vivo measurement of the oxygen saturation of retinal vessels in healthy volunteers," *IEEE Transactions on Biomedical Engineering*, vol. 46, no. 12, pp. 1454–1465, 1999.

[33] F. W. Campbell and W. a. Rushton, "Measurement of the scotopic pigment in the living human eye," *The Journal of Physiology*, vol. 130, pp. 131–147, 1955.

[34] W. A. Rushton, "The difference spectrum and the photosensitivity of rhodopsin in the living human eye," *The Journal of Physiology*, vol. 134, no. 1, pp. 11–29, 1956.

[35] J. I. Morgan and E. N. Pugh, "Scanning laser ophthalmoscope measurement of local fundus reflectance and autofluorescence changes arising from rhodopsin bleaching and regeneration," *Investigative Ophthalmology & Visual Science*, vol. 54, no. 3, pp. 2048–2059, 2013.

[36] T. D. Lamb and E. N. Pugh, "Dark adaptation and the retinoid cycle of vision," *Progress in Retinal and Eye Research*, vol. 23, no. 3, pp. 307–380, 2004.

[37] A. Roorda and D. R. Williams, "The arrangement of the three cone classes in the living human eye," *Nature*, vol. 397, no. 6719, pp. 520–522, 1999.

[38] H. Hofer, J. Carroll, J. Neitz, M. Neitz, and D. R. Williams, "Organization of the human trichromatic cone mosaic," *Journal of Neuroscience*, vol. 25, no. 42, pp. 9669–9679, 2005.

[39] J. J. Hunter, W. H. Merigan, and J. B. Schallek, "Imaging retinal activity in the living eye," *Annual Review of Vision Science*, vol. 5, pp. 15–45, 2019.

[40] S. G. Elner, V. M. Elner, M. G. Field, S. Park, J. R. Heckenlively, and H. R. Petty, "Retinal flavoprotein autofluorescence as a measure of retinal health," *Transactions of the American Ophthalmological Society*, vol. 106, pp. 215–222, 2008.

[41] R. Sharma, D. R. Williams, G. Palczewska, K. Palczewski, and J. J. Hunter, "Two-photon autofluorescence imaging reveals cellular structures throughout the retina of the living primate eye," *Investigative Ophthalmology & Visual Science*, vol. 57, no. 2, pp. 632–646, 2016.

[42] M. G. Field, G. M. Comer, T. Kawaji, H. R. Petty, and V. M. Elner, "Noninvasive imaging of mitochondrial dysfunction in dry age-related macular degeneration," *Ophthalmic Surgery, Lasers and Imaging Retina*, vol. 43, no. 5, pp. 358–365, 2012.

[43] R. Sharma, C. Schwarz, J. J. Hunter, G. Palczewska, K. Palczewski, and D. R. Williams, "Formation and clearance of all-trans-retinol in rods investigated in the living primate eye with two-photon ophthalmoscopy," *Investigative Ophthalmology & Visual Science*, vol. 58, no. 1, pp. 604–613, 2017.

[44] K. Tsunoda, Y. Oguchi, G. Hanazono, and M. Tanifuji, "Mapping cone- and rod-induced retinal responsiveness in macaque retina by optical imaging," *Investigative Ophthalmology & Visual Science*, vol. 45, no. 10, pp. 3820–3826, 2004.

[45] P. J. DeLint, T. T. Berendschot, J. Van De Kraats, and D. Van Norren, "Slow optical changes in human photoreceptors induced by light," *Investigative Ophthalmology & Visual Science*, vol. 41, no. 1, pp. 282–289, 2000.

[46] M. D. Abràmoff, Y. H. Kwon, D. Ts'o, P. Soliz, B. Zimmerman, J. Pokorny, and R. Kardon, "Visual stimulus-induced changes in human near-infrared fundus reflectance," *Investigative Ophthalmology & Visual Science*, vol. 47, no. 2, pp. 715–721, 2006.

[47] G. Hanazono, K. Tsunoda, Y. Kazato, W. Suzuki, and M. Tanifuji, "Functional topography of rod and cone photoreceptors in macaque retina determined by retinal densitometry," *Investigative Ophthalmology & Visual Science*, vol. 53, no. 6, pp. 2796–2803, 2012.

[48] D. Ts'o, J. Schallek, Y. Kwon, R. Kardon, M. Abramoff, and P. Soliz, "Noninvasive functional imaging of the retina reveals outer retinal and hemodynamic intrinsic optical signal origins," *Japanese Journal of Ophthalmology*, vol. 53, no. 4, pp. 334–344, 2009.

[49] J. Schallek, H. Li, R. Kardon, Y. Kwon, M. Abramoff, P. Soliz, and D. Ts'o, "Stimulus-evoked intrinsic optical signals in the retina: Spatial and temporal characteristics," *Investigative Ophthalmology & Visual Science*, vol. 50, no. 10, pp. 4865–4872, 2009.

[50] J. Van de Kraats, T. T. Berendschot, and D. Van Norren, "The pathways of light measured in fundus reflectometry," *Vision Research*, vol. 36, no. 15, pp. 2229–2247, 1996.

[51] K. Grieve and A. Roorda, "Intrinsic signals from human cone photoreceptors," *Investigative Ophthalmology & Visual Science*, vol. 49, no. 2, pp. 713–719, 2008.

[52] R. F. Cooper, W. S. Tuten, A. Dubra, D. H. Brainard, and J. I. W. Morgan, "Non-invasive assessment of human cone photoreceptor function," *Biomedical Optics Express*, vol. 8, no. 11, pp. 5098–5112, 2017.

[53] K. M. Litts, R. F. Cooper, J. L. Duncan, and J. Carroll, "Photoreceptor-based biomarkers in AOSLO retinal imaging," *Investigative Ophthalmology & Visual Science*, vol. 58, no. 6, pp. BIO255–BIO267, 2017.

[54] K. Bizheva, R. Pflug, B. Hermann, B. Považay, H. Sattmann, P. Qiu, E. Anger, H. Reitsamer, S. Popov, J. R. Taylor, A. Unterhuber, P. Ahnelt, and W. Drexler, "Optophysiology: Depth-resolved probing of retinal physiology with functional ultrahigh-resolution optical tomography," *Proceedings of the National Academy of Sciences of the U. S. A.*, vol. 103, no. 13, pp. 5066–5071, 2006.

[55] V. J. Srinivasan, Y. Chen, J. S. Duker, and J. G. Fujimoto, "In vivo functional imaging of intrinsic scattering changes in the human retina with high-speed ultrahigh resolution OCT," *Optics Express*, vol. 17, no. 5, pp. 3861–3877, 2009.

[56] I. Erchova, A. R. Tumlinson, J. Fergusson, N. White, W. Drexler, F. Sengpiel, and J. E. Morgan, "Optophysiological characterisation of inner retina responses with high-resolution optical coherence tomography," *Scientific Reports*, vol. 8, no. 1, pp. 1–12, 2018.

[57] X.-C. Yao and Y.-B. Zhao, "Optical dissection of stimulus-evoked retinal activation," *Optics Express*, vol. 16, no. 17, pp. 12446–12459, 2008.

[58] T. Ling, K. C. Boyle, G. Goetz, P. Zhou, Y. Quan, F. S. Alfonso, T. W. Huang, and D. Palanker, "Full-field interferometric imaging of propagating action potentials," *Light: Science and Applications*, vol. 7, no. 1, pp. 1–5, 2018.

[59] T. Ling, K. C. Boyle, V. Zuckerman, T. Flores, C. Ramakrishnan, K. Deisseroth, and D. Palanker, "High-speed interferometric imaging reveals dynamics of neuronal deformation during the action potential," *Proceedings of the National Academy of Sciences of the U. S. A.*, vol. 117, no. 19, pp. 10278–10285, 2020.

[60] W. Yang and R. Yuste, "In vivo imaging of neural activity," *Nature Methods*, vol. 14, no. 4, pp. 349–359, 2017.

[61] T. Akkin, D. Landowne, and A. Sivaprakasam, "Optical coherence tomography phase measurement of transient changes in squid giant axons during activity," *Journal of Membrane Biology*, vol. 231, no. 1, pp. 35–46, 2009.

[62] O. P. Kocaoglu, Z. Liu, F. Zhang, K. Kurokawa, R. S. Jonnal, and D. T. Miller, "Photoreceptor disc shedding in the living human eye," *Biomedical Optics Express*, vol. 7, no. 11, pp. 4554–4568, 2016.

[63] R. S. Jonnal, O. P. Kocaoglu, Q. Wang, S. Lee, and D. T. Miller, "Phase-sensitive imaging of the outer retina using optical coherence tomography and adaptive optics," *Biomedical Optics Express*, vol. 3, no. 1, pp. 104–124, 2012.

[64] M. Azimipour, J. V. Migacz, R. J. Zawadzki, J. S. Werner, and R. S. Jonnal, "Functional retinal imaging using adaptive optics swept-source OCT at 16 MHz," *Optica*, vol. 6, no. 3, pp. 300–303, 2019.

[65] M. Azimipour, D. Valente, K. V. Vienola, J. S. Werner, R. J. Zawadzki, and R. S. Jonnal, "Optoretinogram: optical measurement of human cone and rod photoreceptor responses to light," *Optics Letters*, vol. 45, no. 17, pp. 4658–4661, 2020.

[66] C. Pfäffle, H. Spahr, D. Hillmann, H. Sudkamp, G. Franke, P. Koch, and G. Hüttmann, "Reduction of frame rate in full-field swept-source optical coherence tomography by numerical motion correction [Invited]," *Biomedical Optics Express*, vol. 8, no. 3, pp. 1499–1511, 2017.

[67] D. Hillmann, T. Bonin, C. Lührs, G. Franke, M. Hagen-Eggert, P. Koch, and G. Hüttmann, "Common approach for compensation of axial motion artifacts in swept-source OCT and dispersion in Fourier-domain oct," *Optics Express*, vol. 20, pp. 6761–6776, 2012.

[68] D. Hillmann, H. Spahr, C. Hain, H. Sudkamp, G. Franke, C. Pfäffle, C. Winter, and G. Hüttmann, "Aberration-free volumetric high-speed imaging of in vivo retina," *Scientific Reports*, vol. 6, pp. 1–11, 2016.

[69] D. Hillmann, C. Pfäffle, H. Spahr, S. Burhan, L. Kutzner, F. Hilge, and G. Hüttmann, "Computational adaptive optics for optical coherence tomography using multiple randomized subaperture correlations," *Optics Letters*, vol. 44, pp. 3905–3908, 2019.

[70] H. Spahr, D. Hillmann, C. Hain, C. Pfäffle, H. Sudkamp, G. Franke, and G. Hüttmann, "Imaging pulse wave propagation in human retinal vessels using full-field swept-source optical coherence tomography," *Optics Letters*, vol. 40, no. 20, pp. 4771–4774, 2015.

[71] J. W. Goodman, *Introduction to Fourier optics*. Roberts & Co. Publishers, 3 ed., 2005.

[72] W. Lauterborn, T. Kurz, and M. Wiesenfeldt, *Kohärente Optik: Grundlagen für Physiker und Ingenieure*. Springer, 1994.

[73] F. Zernike, "Phase contrast, a new method for the microscopic bservation of transparent objects," *Physica*, vol. 9, no. 7, pp. 686 – 698, 1942.

[74] W. L. Bragg, "The diffraction of short electromagnetic waves by a crystal," *Proceedings of the Cambridge Philosophical Society*, vol. 17, pp. 43–57, 1913.

[75] D. Gabor, "Microscopy by reconstructed wave-fronts," *Proceedings of the Royal Society A Mathematical, Physical and Endineering Sciences*, vol. 197, pp. 454–487, 1949.

[76] Y. N. Denisyuk, "On the reflection of optical properties of an object in a wave field of light scattered by it," *Doklady Akademii Nauk SSSR*, vol. 144, no. 6, pp. 1275–1278, 1962.

[77] E. N. Leith and J. Upatnieks, "Reconstructed wavefronts and communication theory," *Journal of the Optical Society of America*, vol. 52, no. 10, pp. 1123–1130, 1962.

[78] J. W. Goodman and R. W. Lawrence, "Digital image formation from electronically detected holograms," *Applied Physics Letters*, vol. 11, no. 3, pp. 77–79, 1967.

[79] D. Huang, E. a. Swanson, C. P. Lin, J. S. Schuman, W. G. Stinson, W. Chang, M. R. Hee, T. Flotire, K. Gregory, C. a. Puliafito, and J. G. Fujimoto, "Optical coherence tomography," vol. 254, pp. 1178–1181, 1991.

[80] M. Wojtkowski, R. Leitgeb, A. Kowalczyk, T. Bajraszewski, and A. F. Fercher, "In vivo human retinal imaging by Fourier domain optical coherence tomography," *Journal of Biomedical Optics*, vol. 7, no. 3, pp. 457–463, 2002.

[81] S. H. Yun, G. J. Tearney, J. F. D. Boer, N. Iftimia, and B. E. Bouma, "High-speed optical frequency-domain imaging," *Optics Express*, vol. 11, no. 22, pp. 2953–2963, 2003.

[82] D. Hillmann, *Holoscopy.* PhD thesis, Universität zu Lübeck, 2014.

[83] A. Dubois, L. Vabre, A. C. Boccara, and E. Beaurepaire, "High-resolution full-field optical coherence tomography with a Linnik microscope," *Applied Optics*, vol. 41, no. 4, pp. 805–812, 2002.

[84] K. Grieve, A. Dubois, M. Simonutti, M. Paques, J. Sahel, J.-F. Le Gargasson, and C. Boccara, "In vivo anterior segment imaging in the rat eye with high speed white light full-field optical coherence tomography," *Optics Express*, vol. 13, no. 16, pp. 6286–6295, 2005.

[85] V. Mazlin, P. Xiao, E. Dalimier, K. Grieve, K. Irsch, J.-A. Sahel, M. Fink, and A. C. Boccara, "In vivo high resolution human corneal imaging using full-field optical coherence tomography," *Biomedical Optics Express*, vol. 9, no. 2, pp. 557–568, 2018.

[86] A. Dubois, K. Grieve, G. Moneron, R. Lecaque, L. Vabre, and C. Boccara, "Ultrahigh-resolution full-field optical coherence tomography," *Applied Optics*, vol. 43, no. 14, pp. 2874–2883, 2004.

[87] J. Scholler, V. Mazlin, O. Thouvenin, K. Groux, P. Xiao, J.-A. Sahel, M. Fink, C. Boccara, and K. Grieve, "Probing dynamic processes in the eye at multiple spatial and temporal scales with multimodal full field OCT," *Biomedical Optics Express*, vol. 10, no. 2, pp. 731–745, 2019.

[88] C. Apelian, F. Harms, O. Thouvenin, and A. C. Boccara, "Dynamic full field optical coherence tomography: subcellular metabolic contrast revealed in tissues by interferometric signals temporal analysis," *Biomedical Optics Express*, vol. 7, no. 4, p. 1511, 2016.

[89] B. Považay, A. Unterhuber, B. Hermann, H. Sattmann, H. Arthaber, and W. Drexler, "Full-field time-encoded frequency-domain optical coherence tomography," *Optics Express*, vol. 14, no. 17, pp. 7661–7669, 2006.

[90] T. Bonin, G. Franke, M. Hagen-Eggert, P. Koch, and G. Hüttmann, "In vivo Fourier-domain full-field oct of the human retina with 15 million A-lines/s," *Optics Letters*, vol. 35, no. 20, pp. 3432–3434, 2010.

[91] S. G. Adie, N. D. Shemonski, B. W. Graf, A. Ahmad, P. Scott Carney, and S. A. Boppart, "Guide-star-based computational adaptive optics for broadband interferometric tomography," *Applied Physics Letters*, vol. 101, no. 22, pp. 221117–1–221117–5, 2012.

[92] V. Matkivsky, A. Moiseev, P. Shilyagin, A. Rodionov, H. Spahr, C. Pfäffle, G. Hüttmann, D. Hillmann, and G. Gelikonov, "Determination and correction of aberrations in full field optical coherence tomography using phase gradient autofocus by maximizing the likelihood function," *Journal of Biophotonics*, vol. 13, p. e202000112, aug 2020.

[93] A. Kumar, T. Kamali, R. Platzer, A. Unterhuber, W. Drexler, and R. A. Leitgeb, "Anisotropic aberration correction using region of interest based digital adaptive optics in Fourier domain OCT," *Biomedical Optics Express*, vol. 6, no. 4, pp. 1124–1134, 2015.

[94] N. D. Shemonski, F. A. South, Y.-Z. Liu, S. G. Adie, P. S. Carney, and S. A. Boppart, "Computational high-resolution optical imaging of the living human retina," *Nature Photonics*, vol. 9, no. 102, pp. 440–443, 2015.

[95] S. G. Adie, B. W. Graf, A. Ahmad, P. S. Carney, and S. A. Boppart, "Computational adaptive optics for broadband optical interferometric tomography of biological tissue," *Proceedings of the National Academy of Sciences of the U. S. A.*, vol. 109, no. 19, pp. 7175–7180, 2012.

[96] D. Sadava, D. M. Hillis, H. C. Heller, and S. D. Hacker, *Purves Biologie*. Springer Spektrum, 10 ed., 2019.

[97] W. Kahle, M. Frotscher, and F. Schmitz, *Taschenatlas Anatomie, Band 3: Nervensystem und Sinnesorgane*. Georg Thieme Verlag, 12 ed., 2018.

[98] Wikipedia, "Zapfen (Auge) - Wikipedia, die freie Enzyklopädie," 2020. Available at: https://de.wikipedia.org/w/index.php?title=Zapfen_(Auge)&oldid=195448801/, accessed 12.10.2020.

[99] L. T. Sharpe, A. Stockman, W. Jagla, and H. Jägle, "A luminous efficiency function, $V^*(\lambda)$, for daylight adaptation," *Journal of Vision*, vol. 5, no. 11, pp. 948–968, 2005.

[100] N. Yagi, "Structural changes in rod outer segments of frog and mouse after illumination," *Experimental Eye Research*, vol. 116, pp. 395–401, 2013.

[101] M. E. C. Webvision, "Anatomy and physiology of the retina," 2012. Available at: https://webvision.med.utah.edu/book/part-ii-anatomy-and-physiology-of-the-retina/, accessed 12.10.2020.

[102] D. A. Baylor, B. J. Nunn, and J. L. Schnapf, "The photocurrent, noise and spectral sensitivity of rods of the monkey Macaca fascicularis," *The Journal of Physiology*, vol. 357, no. 1, pp. 575–607, 1984.

[103] E. N. Pugh and T. D. Lamb, "Chapter 5 Phototransduction in vertebrate rods and cones: Molecular mechanisms of amplification, recovery and light adaptation," in *Molecular Mechanisms in Visual Transduction* (D. Stavenga, W. DeGrip, and E. Pugh, eds.), vol. 3 of *Handbook of Biological Physics*, pp. 183–255, North-Holland, 2000.

[104] F. I. Hárosi, "Absorption spectra and linear dichroism of some amphibian photoreceptors," *Journal of General Physiology*, vol. 66, no. 3, pp. 357–382, 1975.

[105] I. B. Leskov, V. A. Klenchin, J. W. Handy, G. G. Whitlock, V. I. Govardovskii, M. D. Bownds, T. D. Lamb, E. N. Pugh, and V. Y. Arshavsky, "The gain of rod phototransduction: Reconcillation of biochemical and electrophysiological measurements," *Neuron*, vol. 27, no. 3, pp. 525–537, 2000.

[106] M. Heck and K. P. Hofmann, "Maximal rate and nucleotide dependence of rhodopsin-catalyzed transducin activation: Initial rate analysis based on a double displacement mechanism," *Journal of Biological Chemistry*, vol. 276, no. 13, pp. 10000–10009, 2001.

[107] C. M. Krispel, D. Chen, N. Melling, Y. J. Chen, K. A. Martemyanov, N. Quillinan, V. Y. Arshavsky, T. G. Wensel, C. K. Chen, and M. E. Burns, "RGS expression rate-limits recovery of rod photoresponses," *Neuron*, vol. 51, no. 4, pp. 409–416, 2006.

[108] H. E. Hamm and M. D. Bownds, "Protein complement of rod outer segments of frog retina," *Biochemistry*, vol. 25, no. 16, pp. 4512–4523, 1986.

[109] A. Sitaramayya, J. Harkness, J. H. Parkes, C. Gonzalez-Oliva, and P. A. Liebman, "Kinetic studies suggest that light-activated cyclic GMP phosphodiesterase is a complex with G-protein subunits," *Biochemistry*, vol. 25, no. 3, pp. 651–656, 1986.

[110] J. S. Anant, O. C. Ong, H. Xie, S. Clarke, P. J. O'Brien, and B. K. Fung, "In vivo differential prenylation of retinal cyclic GMP phosphodiesterase catalytic subunits," *Journal of Biological Chemistry*, vol. 267, no. 2, pp. 687–690, 1992.

[111] W. Baehr, M. J. Devlin, and M. L. Applebury, "Isolation and characterization of cGMP phosphodiesterase from bovine rod outer segments," *Journal of Biological Chemistry*, vol. 254, no. 22, pp. 11669–11677, 1979.

[112] K. Yau, "Cyclic GMP-activated conductance of retinal photoreceptor cells," *Annual Review of Neuroscience*, vol. 12, no. 1, pp. 289–327, 1989.

[113] J. W. Karpen, A. L. Zimmerman, L. Stryer, and D. A. Baylor, "Gating kinetics of the cyclic-GMP-activated channel of retinal rods: Flash photolysis and voltage-jump studies," *Proceedings of the National Academy of Sciences of the U. S. A.*, vol. 85, no. 4, pp. 1287–1291, 1988.

[114] K. W. Yau, "Phototransduction mechanism in retinal rods and cones," *Investigative Ophthalmology & Visual Science*, vol. 35, no. 1, pp. 9–32, 1994.

[115] W. A. Hagins, R. D. Penn, and S. Yoshikami, "Dark current and photocurrent in retinal rods," *Biophysical Journal*, vol. 10, no. 5, pp. 380–412, 1970.

[116] T. W. Kraft, D. M. Schneeweis, and J. L. Schnapf, "Visual transduction in human rod photoreceptors," *Journal of Physiology*, vol. 464, no. 1, pp. 747–765, 1993.

[117] H. Wässle, C. Puller, F. Müller, and S. Haverkamp, "Cone contacts, mosaics, and territories of bipolar cells in the mouse retina," *Journal of Neuroscience*, vol. 29, no. 1, pp. 106–117, 2009.

[118] H. Wässle, "Parallel processing in the mammalian retina," *Nature Reviews Neuroscience*, vol. 5, no. 10, pp. 747–757, 2004.

[119] T. Euler, S. Haverkamp, T. Schubert, and T. Baden, "Retinal bipolar cells: Elementary building blocks of vision," *Nature Reviews Neuroscience*, vol. 15, no. 8, pp. 507–519, 2014.

[120] S. A. Bloomfield and R. F. Dacheux, "Rod vision: Pathways and processing in the mammalian retina," *Progress in Retinal and Eye Research*, vol. 20, no. 3, pp. 351–384, 2001.

[121] C. Puller and S. Haverkamp, "Bipolar cell pathways for color vision in non-primate dichromats," *Visual Neuroscience*, vol. 28, no. 1, pp. 51–60, 2011.

[122] T. Breuninger, C. Puller, S. Haverkamp, and T. Euler, "Chromatic bipolar cell pathways in the mouse retina," *Journal of Neuroscience*, vol. 31, no. 17, pp. 6504–6517, 2011.

[123] W. Li and S. H. DeVries, "Bipolar cell pathways for color and luminance vision in a dichromatic mammalian retina," *Nature Neuroscience*, vol. 9, no. 5, pp. 669–675, 2006.

[124] E. Hartveit, "Functional organization of cone bipolar cells in the rat retina," *Journal of Neurophysiology*, vol. 77, no. 4, pp. 1716–1730, 1997.

[125] R. H. Masland, "The fundamental plan of the retina," *Nature Neuroscience*, vol. 4, no. 9, pp. 877–886, 2001.

[126] S. Cajal, "La retine des vertebres," in *The vertebrate retina*, 1893.

[127] R. H. Masland, "The neuronal organization of the retina," *Neuron*, vol. 76, no. 2, pp. 266–280, 2012.

[128] J. R. Sanes and R. H. Masland, "The types of retinal ganglion cells: current status and implications for neuronal classification," *Annual Review of Neuroscience*, vol. 38, no. 1, pp. 221–246, 2015.

[129] H. B. Barlow, R. M. Hill, and W. R. Levick, "Retinal ganglion cells responding selectively to direction and speed of image motion in the rabbit," *The Journal of Physiology*, vol. 173, no. 3, pp. 377–407, 1964.

[130] H. B. Barlow and W. R. Levick, "The mechanism of directionally selective units in rabbit's retina.," *The Journal of Physiology*, vol. 178, no. 3, pp. 477–504, 1965.

[131] S. Weng, W. Sun, and S. He, "Identification of ON-OFF direction-selective ganglion cells in the mouse retina," *Journal of Physiology*, vol. 562, no. 3, pp. 915–923, 2005.

[132] C. W. Oyster and H. B. Barlow, "Direction-selective units in rabbit retina: Distribution of preferred directions," *Science*, vol. 155, no. 3764, pp. 841–842, 1967.

[133] J. Elstrott, A. Anishchenko, M. Greschner, A. Sher, A. M. Litke, E. J. Chichilnisky, and M. B. Feller, "Direction selectivity in the retina is established independent of visual experience and cholinergic retinal waves," *Neuron*, vol. 58, no. 4, pp. 499–506, 2008.

[134] S. Hattar, H. W. Liao, M. Takao, D. M. Berson, and K. W. Yau, "Melanopsin-containing retinal ganglion cells: Architecture, projections, and intrinsic photosensitivity," *Science*, vol. 295, no. 5557, pp. 1065–1070, 2002.

[135] S. Hattar, R. J. Lucas, N. Mrosovsky, S. Thompson, R. H. Douglas, M. W. Hankins, J. Lem, M. Biel, F. Hofmann, R. G. Foster, and K. W. Yau, "Melanopsin and rod – cone photoreceptive systems account for all major accessory visual functions in mice," *Nature*, vol. 424, no. 6944, pp. 76–81, 2003.

[136] S. Panda, T. K. Sato, A. M. Castrucci, M. D. Rollag, W. J. DeGrip, J. B. Hogenesch, I. Provencio, and S. A. Kay, "Melanopsin (Opn4) requirement for normal light-induced circadian phase shifting," *Science*, vol. 298, no. 5601, pp. 2213–2216, 2002.

[137] X. Qiu, T. Kumbalasiri, S. M. Carlson, K. Y. Wong, V. Krishna, I. Provencio, and D. M. Berson, "Induction of photosensitivity by heterologous expression of melanopsin," *Nature*, vol. 433, no. 7027, pp. 745–749, 2005.

[138] S. A. Bloomfield, "Orientation-sensitive amacrine and ganglion cells in the rabbit retina," *Journal of Neurophysiology*, vol. 71, no. 5, pp. 1672–1691, 1994.

[139] H. Shigang, W. R. Levick, and D. I. Vaney, "Distinguishing direction selectivity from orientation selectivity in the rabbit retina," *Visual Neuroscience*, vol. 15, no. 3, pp. 439–447, 1998.

[140] B. Cleland and W. R. Levick, "Proberties of rarely encoutered types of ganglion cells in the cat's retina and an overall classification," *Journal of Physiology*, vol. 240, pp. 457–492, 1974.

[141] G. H. Jacobs and R. B. Tootell, "Spectrally-opponent responses in ground squirrel optic nerve," *Vision Research*, vol. 20, no. 1, pp. 9–13, 1980.

[142] J. M. Hemmi, A. James, and W. R. Taylor, "Color opponent retinal ganglion cells in the tammar wallaby retina," *Journal of Vision*, vol. 2, no. 9, pp. 608–617, 2002.

[143] S. Chen and W. Li, "A color-coding amacrine cell may provide a blue-Off signal in a mammalian retina," *Nature Neuroscience*, vol. 15, no. 7, pp. 954–956, 2012.

[144] A. Sher and S. H. Devries, "A non-canonical pathway for mammalian blue-green color vision," *Nature Neuroscience*, vol. 15, no. 7, pp. 952–953, 2012.

[145] L. Yin, R. G. Smith, P. Sterling, and D. H. Brainard, "Physiology and morphology of color-opponent ganglion cells in a retina expressing a dual gradient of S and M opsins," *Journal of Neuroscience*, vol. 29, no. 9, pp. 2706–2724, 2009.

[146] T. Okano, D. Kojima, Y. Fukada, Y. Shichida, and T. Yoshizawa, "Primary structures of chicken cone visual pigments: Vertebrate rhodopsins have evolved out of cone visual pigments," *Proceedings of the National Academy of Sciences of the U. S. A.*, vol. 89, no. 13, pp. 5932–5936, 1992.

[147] R. L. Johnson, K. B. Grant, T. C. Zankel, M. F. Boehm, K. Nakanishi, S. L. Merbs, J. Nathans, and J. Nathans, "Cloning and expression of goldfish opsin sequences," *Biochemistry*, vol. 32, no. 1, pp. 208–214, 1993.

[148] M. Van Wyk, W. R. Taylor, and D. I. Vaney, "Local edge detectors: A substrate for fine spatial vision at low temporal frequencies in rabbit retina," *Journal of Neuroscience*, vol. 26, no. 51, pp. 13250–13263, 2006.

[149] S. A. Baccus, B. P. Ölveczky, M. Manu, and M. Meister, "A retinal circuit that computes object motion," *Journal of Neuroscience*, vol. 28, no. 27, pp. 6807–6817, 2008.

[150] T. L. Russell and F. S. Werblin, "Retinal synaptic pathways underlying the response of the rabbit local edge detector," *Journal of Neurophysiology*, vol. 103, no. 5, pp. 2757–2769, 2010.

[151] F. R. Amthor, E. S. Takahashi, and C. W. Oyster, "Morphologies of rabbit retinal ganglion cells with concentric receptive fields," *Journal of Comparative Neurology*, vol. 280, no. 1, pp. 72–96, 1989.

[152] D. I. Vaney, B. Sivyer, and W. R. Taylor, "Direction selectivity in the retina: Symmetry and asymmetry in structure and function," *Nature Reviews Neuroscience*, vol. 13, no. 3, pp. 194–208, 2012.

[153] Y. Zhang, I. J. Kim, J. R. Sanes, and M. Meister, "The most numerous ganglion cell type of the mouse retina is a selective feature detector," *Proceedings of the National Academy of Sciences of the U. S. A.*, vol. 109, no. 36, pp. E2391–E2398, 2012.

[154] I. J. Kim, Y. Zhang, M. Meister, and J. R. Sanes, "Laminar restriction of retinal ganglion cell dendrites and axons: Subtype-specific developmental patterns revealed with transgenic markers," *Journal of Neuroscience*, vol. 30, no. 4, pp. 1452–1462, 2010.

[155] W. Levick, "Receptive fields and trigger features of ganglion cells in the visual streak of the rabbit's retina," *Journal of physiology*, vol. 188, pp. 285–307, 1967.

[156] H. Spahr, *Detektion von retinalen Mikrobewegungen — Anwendungen der phasensensitiven optischen Kohärenztomografie.* PhD thesis, Universität zu Lübeck, 2018.

[157] S. Martinez-Conde, S. L. Macknik, and D. H. Hubel, "The role of fixational eye movements in visual perception," *Nature Reviews Neuroscience*, vol. 5, no. 3, pp. 229–240, 2004.

[158] S. Martinez-Conde and S. L. Macknik, "Fixational eye movements across vertebrates: Comparative dynamics, physiology, and perception," *Journal of Vision*, vol. 8, no. 14, pp. 1–16, 2008.

[159] M. R. Hee, J. A. Izatt, E. A. Swanson, D. Huang, J. S. Schuman, C. P. Lin, C. A. Puliafito, and J. G. Fujimoto, "Optical coherence tomography of the human retina," *Archives of Ophthalmology*, vol. 113, no. 3, pp. 325–332, 1995.

[160] Y. Huang, A. V. Cideciyan, G. I. Papastergiou, E. Banin, S. L. Semple-Rowland, A. H. Milam, and S. G. Jacobson, "Relation of optical coherence tomography to microanatomy in normal and rd chickens," *Investigative Ophthalmology & Visual Science*, vol. 39, no. 12, pp. 2405–2416, 1998.

[161] A. Herzog, K. L. Boyer, and C. Roberts, "Robust extraction of the optic nerve head in optical coherence tomography," in *Computer Vision and Mathematical Methods in Medical and Biomedical Image Analysis* (M. Sonka, I. A. Kakadiaris, and J. Kybic, eds.), pp. 395–407, Springer Berlin Heidelberg, 2004.

[162] K. A. Vermeer, J. van der Schoot, H. G. Lemij, and J. F. de Boer, "Automated segmentation by pixel classification of retinal layers in ophthalmic OCT images," *Biomedical Optics Express*, vol. 2, no. 6, pp. 1743–1756, 2011.

[163] A. Fuller, R. Zawadzki, S. Choi, D. Wiley, J. Werner, and B. Hamann, "Segmentation of three-dimensional retinal image data," *IEEE Transactions on Visualization and Computer Graphics*, vol. 13, no. 6, pp. 1719–1726, 2007.

[164] M. A. Mayer, R. P. Tornow, R. Bock, J. Hornegger, and F. E. Kruse, "Automatic nerve fiber layer segmentation and geometry correction on spectral domain OCT images ssing fuzzy C-means clustering," *Investigative Ophthalmology & Visual Science*, vol. 49, no. 13, p. 1880, 2008.

[165] D. C. Fernández, "Delineating fluid-filled region boundaries in optical coherence tomography images of the retina," *IEEE Transactions on Medical Imaging*, vol. 24, no. 8, pp. 929–945, 2005.

[166] M. Mujat, R. C. Chan, B. Cense, B. Hyle Park, C. Joo, T. Akkin, T. C. Chen, and J. F. de Boer, "Retinal nerve fiber layer thickness map determined from optical coherence tomography images," *Optics Express*, vol. 13, no. 23, pp. 9480–9491, 2005.

[167] A. Yazdanpanah, G. Hamarneh, B. Smith, and M. Sarunic, "Intra-retinal layer segmentation in optical coherence tomography using an active contour approach," in *Medical Image Computing and Computer-Assisted Intervention – MICCAI 2009* (G.-Z. Yang, D. Hawkes, D. Rueckert, A. Noble, and C. Taylor, eds.), pp. 649–656, Springer Berlin Heidelberg, 2009.

[168] A. Mishra, A. Wong, K. Bizheva, and D. A. Clausi, "Intra-retinal layer segmentation in optical coherence tomography images," *Optics Express*, vol. 17, no. 26, pp. 23719–23728, 2009.

[169] D. A. Tolliver, I. Koutis, H. Ishikawa, J. S. Schuman, and G. L. Miller, "Automatic multiple retinal layer segmentation in spectral domain OCT scans via spectral rounding," *Investigative Ophthalmology & Visual Science*, vol. 49, no. 13, p. 1878, 2008.

[170] R. Kafieh, H. Rabbani, M. Abramoff, and M. Sonka, "Intra-retinal layer segmentation of optical coherence tomography using diffusion map," *ICASSP, IEEE International Conference on Acoustics, Speech and Signal Processing - Proceedings*, vol. 17, no. 8, pp. 1080–1084, 2013.

[171] M. D. Abràmoff, K. Lee, M. Niemeijer, W. L. Alward, E. C. Greenlee, M. K. Garvin, M. Sonka, and Y. H. Kwon, "Automated segmentation of the cup and rim from spectral domain OCT of the optic nerve head," *Investigative Ophthalmology & Visual Science*, vol. 50, no. 12, pp. 5778–5784, 2009.

[172] M. K. Garvin, M. D. Abramoff, R. Kardon, S. R. Russell, X. Wu, and M. Sonka, "Intraretinal layer segmentation of macular optical coherence tomography images using optimal 3-D graph search," *IEEE Transactions on Medical Imaging*, vol. 27, no. 10, pp. 1495–1505, 2008.

[173] M. Haeker, M. Abràmoff, R. Kardon, and M. Sonka, "Segmentation of the surfaces of the retinal layer from OCT images," in *Medical Image Computing and Computer-Assisted Intervention – MICCAI 2006* (R. Larsen, M. Nielsen, and J. Sporring, eds.), (Berlin, Heidelberg), pp. 800–807, Springer Berlin Heidelberg, 2006.

[174] R. R. Coifman and S. Lafon, "Diffusion maps," *Applied and Computational Harmonic Analysis*, vol. 21, no. 1, pp. 5–30, 2006.

[175] B. Nadler, S. Lafon, R. R. Coifman, and I. G. Kevrekidis, "Diffusion maps, spectral clustering and reaction coordinates of dynamical systems," *Applied and Computational Harmonic Analysis*, vol. 21, no. 1, pp. 113–127, 2006.

[176] B. Nadler, S. Lafon, R. Coifman, and I. Kevrekidis, "Diffusion maps, spectral clustering and eigenfunctions of fokker-planck operators," in *NIPS*, 2005.

[177] S. Lloyd, "Least squares quantization in PCM," *IEEE Transactions on Information Theory*, vol. 28, no. 2, pp. 129–137, 1982.

[178] J. C. Bezdek, "Objective function clustering," in *Pattern Recognition with Fuzzy Objective Function Algorithms*, pp. 43–93, Springer, Boston, MA, 1981.

[179] A. P. Dempster, N. M. Laird, and D. B. Rubin, "Maximum likelihood from incomplete data via the EM algorithm," *Journal of the Royal Statistical Society: Series B (Methodological)*, vol. 39, no. 1, pp. 1–22, 1977.

[180] M. Daszykowski and B. Walczak, "Density-based clustering methods," *Comprehensive Chemometrics*, vol. 2, pp. 635–654, 2009.

[181] M. Ankerst, M. M. Breunig, H.-P. Kriegel, and J. Sander, "OPTICS: Ordering points to identify the clustering structure," in *ACM SIGMOD Record*, vol. 28, pp. 49–60, ACM Press, 1999.

[182] L. Xu, J. Neufeld, B. Larson, and D. Schuurmans, "Maximum margin clustering," in *Advances in Neural Information Processing Systems* (L. Saul, Y. Weiss, and L. Bottou, eds.), vol. 17, pp. 1537–1544, MIT- Press, 2005.

[183] K. Gercke, *Optimization and Evaluation of a Segmentation Algorithm for Full-Field Swept-Source Optical Coherence Tomography.* Bachelor's thesis at Universität zu Lübeck, 2019.

[184] K. T. Knox and B. J. Thompson, “Recovery of images from atmospherically degraded short-exposure photographs,” *Astrophysical Journal*, vol. 193, p. L45, 1974.

[185] G. R. Ayers, M. J. Northcott, and J. C. Dainty, “Knox-Thompson and triple-correlation imaging through atmospheric turbulence,” *Journal of the Optical Society of America A*, vol. 5, no. 7, pp. 963–984, 1988.

[186] J. C. Marron, P. P. Sanchez, and R. C. Sullivan, “Unwrapping algorithm for least-squares phase recovery from the modulo 2π bispectrum phase,” *Journal of the Optical Society of America A*, vol. 7, no. 1, pp. 14–20, 1990.

[187] A. W. Lohmann, G. Weigelt, and B. Wirnitzer, “Speckle masking in astronomy: triple correlation theory and applications,” *Applied Optics*, vol. 22, no. 24, pp. 4028–4037, 1983.

[188] A. Lannes, “Backprojection mechanisms in phase-closure imaging. Bispectral analysis of the phase-restoration process,” *Experimental Astronomy*, vol. 1, no. 1, pp. 47–76, 1989.

[189] H. Spahr, C. Pfäffle, S. Burhan, L. Kutzner, F. Hilge, G. Hüttmann, and D. Hillmann, “Phase-sensitive interferometry of decorrelated speckle patterns,” *Scientific Reports*, vol. 9, no. 1, p. 11748, 2019.

[190] D. Hillmann, H. Spahr, C. Pfäffle, H. Sudkamp, G. Franke, and G. Hüttmann, “In vivo optical imaging of physiological responses to photostimulation in human photoreceptors,” *Proceedings of the National Academy of Sciences of the U. S. A.*, vol. 113, no. 46, pp. 13138–13143, 2016.

[191] A. J. Zele and D. Cao, “Vision under mesopic and scotopic illumination,” *Frontiers in Psychology*, vol. 5, no. 1, pp. 1–15, 2015.

[192] M. D. Abràmoff, R. F. Mullins, K. Lee, J. M. Hoffmann, M. Sonka, D. B. Critser, S. F. Stasheff, and E. M. Stone, “Human photoreceptor outer segments shorten during light adaptation,” *Investigative Ophthalmology & Visual Science*, vol. 54, no. 5, pp. 3721–3728, 2013.

[193] R. Nelson, E. V. Famiglietti, and H. Kolb, “Intracellular staining reveals different levels of stratification for on- and off-center ganglion cells in cat retina,” *Journal of Neurophysiology*, vol. 41, no. 2, pp. 472–83, 1978.

[194] T. Yeh, B. B. Lee, and J. Kremers, “Temporal response of ganglion cells of the macaque retina to cone-specific modulation,” vol. 12, no. 3, pp. 456–464, 1995.

[195] B. B. Lee, P. R. Martin, and A. Valbergt, "Sensitivity of macaque retinal ganglion cells to chromatic and luminance flicker," *Journal of Physiology*, vol. 414, pp. 223–243, 1989.

[196] C. Pfäffle, H. Spahr, L. Kutzner, S. Burhan, F. Hilge, Y. Miura, G. Hüttmann, and D. Hillmann, "Simultaneous functional imaging of neuronal and photoreceptor layers in living human retina," *Optics Letters*, vol. 44, no. 23, pp. 5671–5674, 2019.

[197] N. Drasdo, C. L. Millican, C. R. Katholi, and C. A. Curcio, "The length of Henle fibers in the human retina and a model of ganglion receptive field density in the visual field," *Vision Research*, vol. 47, no. 22, pp. 2901–2911, 2007.

[198] F. W. Campbell and W. A. H. Rushton, "Measurement of the scotopic pigment in the living human eye.," *The Journal of Physiology*, vol. 130, pp. 131–147, 1955.

[199] M. Azimipour, J. V. Migacz, R. J. Zawadzki, J. S. Werner, and R. S. Jonnal, "Functional retinal imaging using adaptive optics swept-source OCT at 16 MHz," *Optica*, vol. 6, no. 3, pp. 300–303, 2019.

[200] O. P. Kocaoglu, T. L. Turner, Z. Liu, and D. T. Miller, "Adaptive optics optical coherence tomography at 1 MHz," *Biomedical Optics Express*, vol. 5, no. 12, pp. 4186–4200, 2014.

[201] V. P. Pandiyan, A. Maloney-Bertelli, J. A. Kuchenbecker, K. C. Boyle, T. Ling, Z. C. Chen, B. H. Park, A. Roorda, D. Palanker, and R. Sabesan, "The optoretinogram reveals the primary steps of phototransduction in the living human eye," *Science Advances*, vol. 6, no. 37, p. eabc1124, 2020.

[202] F. Zhang, K. Kurokawa, A. Lassoued, J. A. Crowell, and D. T. Miller, "Cone photoreceptor classification in the living human eye from photostimulation-induced phase dynamics," *Proceedings of the National Academy of Sciences of the U. S. A.*, vol. 116, no. 16, pp. 7951–7956, 2019.

[203] T. Wilson and C. J. R. Shepparu, *Theory and practice of scanning optical microscopy.* Academic Press, London, 1984.

[204] T. Berlind, G. K. Pribil, D. Thompson, J. A. Woollam, and H. Arwin, "Effects of ion concentration on refractive indices of fluids measured by the minimum deviation technique," *Physica Status Solidi (C) Current Topics in Solid State Physics*, vol. 5, no. 5, pp. 1249–1252, 2008.

[205] C. Bardy, M. van den Hurk, T. Eames, C. Marchand, R. V. Hernandez, M. Kellogg, M. Gorris, B. Galet, V. Palomares, J. Brown, A. G. Bang, J. Mertens, L. Böhnke, L. Boyer, S. Simon, and F. H. Gage, "Neuronal medium that supports basic synaptic functions and activity of human neurons in vitro.," *Proceedings of the National Academy of Sciences of the U. S. A.*, vol. 112, no. 20, pp. E2725–34, 2015.

[206] P. Zhang, R. J. Zawadzki, M. Goswami, P. T. Nguyen, V. Yarov-Yarovoy, M. E. Burns, and E. N. Pugh, "In vivo optophysiology reveals that G-protein activation triggers osmotic swelling and increased light scattering of rod photoreceptors," *Proceedings of the National Academy of Sciences of the U. S. A.*, vol. 114, no. 14, pp. E2937–E2946, 2017.

[207] A. V. Dmitriev, V. I. Govardovskii, H. N. Schwahn, and R. H. Steinberg, "Light-induced changes of extracellular ions and volume in the isolated chick retina-pigment epithelium preparation," *Visual Neuroscience*, vol. 16, no. 6, pp. 1157–1167, 1999.

[208] B. Huang and C. Karwoski, "Light-evoked expansion of subretinal space in the retina," *Jounral of Neurosience*, vol. 12, pp. 4243–4252, 1992.

[209] J. Li, R. Gallemore, A. Dmitriev, and R. H. Steinberg, "Light-dependent hydration of the space surrounding phototreceptors in chicken retina," *Investigative Ophthalmology & Visual Science*, vol. 35, pp. 2700–2711, 1994.

[210] J. L. van Hemmen and C. Leibold, "Elementary excitations of biomembranes: Differential geometry of undulations in elastic surfaces," *Physics Reports*, vol. 444, no. 2, pp. 51–99, 2007.

[211] K. Olbrich, W. Rawicz, D. Needham, and E. Evans, "Water permeability and mechanical strength of polyunsaturated lipid bilayers," *Biophysical Journal*, vol. 79, no. 1, pp. 321–327, 2000.

[212] H. W. Lim, M. Wortis, and R. Mukhopadhyay, "Stomatocyte-discocyte-echinocyte sequence of the human red blood cell: Evidence for the bilayer-couple hypothesis from membrane mechanics," *Proceedings of the National Academy of Sciences of the U. S. A.*, vol. 99, no. 26, pp. 16766–16769, 2002.

[213] R. Mukhopadhyay, H. W. Gerald Lim, and M. Wortis, "Echinocyte shapes: Bending, stretching, and shear determine spicule shape and spacing," *Biophysical Journal*, vol. 82, no. 4, pp. 1756–1772, 2002.

[214] A. I. Cohen, "Electron microscope observations on form changes in photoreceptor outer segments and their saccules in response to osmotic stress," *Journal of Cell Biology*, vol. 48, no. 3, pp. 547–565, 1971.

[215] Y. Yang, X.-W. Liu, H. Yu, Y. Guan, S. Wang, and N. Tao, "Imaging action potential in single mammalian neurons by tracking the accompanying subnanometer mechanical motion," *ACS Nano*, vol. 12, no. 5, pp. 4186–4193, 2018.

[216] S. Oh, C. Fang-Yen, W. Choi, Z. Yaqoob, D. Fu, Y. Park, R. R. Dassari, and M. S. Feld, "Label-free imaging of membrane potential using membrane electromotility," *Biophysical Journal*, vol. 103, no. 1, pp. 11–18, 2012.

[217] B. Hill, E. Schubert, M. Nokes, and R. Michelson, "Laser interferometer measurement of changes in crayfish axon diameter concurrent with action potential," *Science*, vol. 196, no. 4288, pp. 426–428, 1977.

[218] K. Nakatani and K.-W. Yau, "Calcium and light adaption in retinal rods and cones," *Nature*, vol. 334, no. 7, pp. 8–10, 1988.

[219] K. Pagh-Roehl, E. Wang, and B. Burnside, "Shortening of the calycal process actin cytoskeleton is correlated with myoid elongation in teleost rods," *Experimental Eye Research*, vol. 55, pp. 735–746, 1992.

[220] Z. Liu, K. Kurokawa, F. Zhang, J. J. Lee, and D. T. Miller, "Imaging and quantifying ganglion cells and other transparent neurons in the living human retina," *Proceedings of the National Academy of Sciences of the U. S. A.*, vol. 114, no. 48, pp. 12803–12808, 2017.

[221] E. Auksorius, D. Borycki, and M. Wojtkowski, "Crosstalk-free volumetric in vivo imaging of a human retina with Fourier-domain full-field optical coherence tomography," *Biomedical Optics Express*, vol. 10, no. 12, pp. 6390–6407, 2019.

[222] D. Borycki, M. Hamkało, M. Nowakowski, M. Szkulmowski, and M. Wojtkowski, "Spatiotemporal optical coherence (STOC) manipulation suppresses coherent cross-talk in full-field swept-source optical coherence tomography," *Biomedical Optics Express*, vol. 10, no. 4, pp. 2032–2054, 2019.

[223] B. Kabuth, *Investigation of Different Nutrient Media for Porcine Retina to Optimize the Cell Vitality of Photoreceptors.* Bachelor's thesis at Universität zu Lübeck, 2018.

[224] S. Burhan, "Osmotic analysis and functional imaging of ex-vivo porcine retina using full-field swept-source optical coherence tomography," Master's thesis, Universität zu Lübeck, 2019.

Danksagung

Vor und während der Anfertigung meiner Dissertation hatte ich das Glück mit einer Vielzahl von Menschen zusammenzuarbeiten, welche mir durch ihre fachliche und menschliche Unterstützung auf sehr unterschiedliche Weise beim Erstellen dieser Arbeit geholfen haben, weswegen ich mich an dieser Stelle bedanken möchte.

Zunächst möchte ich mich bei dem ehemaligen und derzeitigen Leiter des Institutes für Biomedizinische Optik, Alfred Vogel und Robert Huber, für die Möglichkeit bedanken diese Arbeit an ihrem Institut durchführen zu können.

Ein ganz besonderer Dank gilt meinem Betreuer und Doktorvater Gereon Hüttmann, der wesentlich zum Erfolg dieser Arbeit beigetragen hat. Durch seinen fachlichen Beistand in Form von kontinuierlichen Diskussionen, Anregungen und neuen Ideen hat er maßgeblich zur Qualität dieser Arbeit beigetragen. Sein hohes Engagement, seine persönliche Begeisterung und sein Einsatz für seine Projekte aber vor allem auch für seine Mitarbeiter haben dazu geführt, dass ich während der letzten Jahre nie die Begeisterung und Motivation für die Arbeit in der funktionellen OCT verloren habe. Außerdem hat er damit ein Arbeitsklima geschaffen, dass man nur sehr ungern wieder verlassen möchte.

Dem Betreuer meiner Masterarbeit Hendrik Spahr möchte ich dafür danken, dass er mir jede dumme Frage die ich zu OCT hatte mit einer bewundernswerten Geduld bis ins letzte Detail beantwortet hat. Dabei hatte die Ausbildung neuer Studenten bei ihm immer eine der höchsten Prioritäten, weswegen er sich auch in stressigen Phasen immer Zeit genommen hat mir alle nur erdenklichen Zusammenhänge und Hintergründe mit einem anscheinend ;) unerschöpflichen Fachwissen zu erklären. Weiter möchte ich mich bei Hendrik Spahr für die Zusammenarbeit während meiner Dissertation bedanken; für die unzähligen hilfreichen Diskussionen aber vor allem dafür, dass es immer selbstverständlich war sich gegenseitig auszuhelfen und dass man sich über die Erfolge der anderen genauso freuen konnte wie für seine eigenen.

Des Weiteren möchte ich mich bei Dierck Hillmann bedanken, der in vielerlei Hinsicht die Grundlagen für die hier vorliegende Arbeit geschaffen hat. Nicht nur ist der hier verwendete Holoscopy Aufbau sowie ein Teil des Post-Processings während seiner Dissertation und darüber hinaus von ihm entwickelt worden, auch stammt die initiale Idee für funktionelle Bildgebung mittels Phasenkontrast von ihm. Ich möchte mich dafür bedanken, dass er sich alle nur erdenkliche Zeit genommen hat, mich in all das einzuarbeiten und für die vielen, vielen, vielen Diskussionen, die ich mit ihm über die letzten Jahre hatte, aus denen eine unfassbare Menge an klugen Ideen entstanden sind, die maßgeblich die Qualität dieser Arbeit beeinflusst haben. Vor allem aber möchte ich mich dafür bedanken, dass ich mit ihm immer einen guten Freund an meiner Seite hatte, auf den ich mich verlassen konnte, wenn es Mal nicht so lief wie ich mir das gewünscht hatte.

Zusätzlich möchte ich allen Danken, die vor mir an dem FF-SS-OCT Aufbau gearbeitet haben und durch ihre Vorarbeit es erst ermöglicht haben, dass ich die hier vorliegende Arbeit in der Form erstellen konnte. Besonders möchte ich dabei Carola Lührs und Helge Sudkamp danken, die unmittelbar vor mir an dem FF-SS-OCT Aufbau gearbeitet haben und mich dementsprechend in alle Kniffe und Geheimnisse des Aufbaus einweihen konnten. Zusätzlich möchte ich Carola Lührs noch dafür danken, dass sie so hohe Maßstäbe an die Laborbuchführung etabliert hat.

Weiter möchte ich Sabrina Lohmann, Bastian Kabuth, Sazan Burhan, Felix Hilge, Lisa Kutzner, Katharina Gercke, Salim Melikov, Jonas Franke, David Mellenberg, Svea Höhl und Léo Puyo dafür danken, dass jeder einzelne von ihnen in den letzten Jahren zu einer entspannten und dennoch produktiven Arbeitsatmosphäre beigetragen hat.

Bei Lisa Kutzner, Sazan Burhan, Katharina Gercke und David Mellenberg möchte ich mich dafür bedanken, dass sie tagelang mit mir in einem dunklen Optiklabor mit stupider Arbeit zugebracht haben um die Messreihen dieser Arbeit zu generieren.

Ich möchte mich bei Bastian Kabuth, Sazan Burhan, Lisa Kutzner und Katharina Gercke dafür bedanken, dass sie stillschweigend meine ersten Gehversuche als Betreuer ertragen haben und sich nicht beschwert haben, wenn ihre Fragen zuerst mit inkohärenten Satzfragmenten beantwortet wurden.

Yoko Miura möchte ich für ihre Zusammenarbeit bezüglich der funktionellen Bildgebung in ex-vivio Retinapräparaten danken, sowie für ihren hohen Einsatz und die große Leidenschaft, mit denen sie sich diesen Projekten gewidmet hat und dafür dass sie bei allen klinischen und biologischen Fragen, die während der letzten Jahre aufgekommen sind, einen schnellen und sehr ausführlichen Rat parat hatte.

Ich möchte Ralf Brinkmann und dem Medizinischen Laserzentrum Lübeck, dafür danken, dass sie mich in den ersten Monaten der Promotion durch ein Stipendium finanziert haben. Kevin Fischer möchte ich dafür danken, dass er sich so schnell und unkompliziert um die Realisierung dieses Stipendiums gekümmert hat.

Zum Schluss möchte ich meinen Eltern Isis und Roland Pfäffle danken, die mir die bestmögliche finanzielle, intellektuelle und vor allem emotionale Unterstützung gegeben haben die man sich wünschen kann. Sie haben mir damit ein Leben ermöglicht, in dem mir alle nur erdenklichen Wege offen standen und die Freiheit gegeben aus all diesen Wegen meinen eigenen zu wählen.

Infinite Science Publishing